THE OLYMPIC RAIN FOREST

An Ecological Web

Ruth Kirk
with Jerry Franklin

Foreword by Ivan Doig
Photography by Ruth and Louis Kirk

University of Washington Press
Seattle and London

Updated edition, 2001
15 14 13 12 11 10 09 8 7 6 5 4

University of Washington Press
PO Box 50096
Seattle, WA 98145-5096
www.washington.edu/uwpress

Library of Congress Cataloging-in-Publication Data

Kirk, Ruth.
The Olympic rain forest : an ecological web / Ruth Kirk with Jerry Franklin : foreword by Ivan Doig.
p. cm.
Includes bibliographical references and index.
ISBN 978-0-295-97187-2 (pbk. : alk. paper)
1. Rain forest ecology—Washington (State)—Olympic Peninsula. 2. Natural history—Washington (State)—Olympic Peninsula. 3. Rain forest ecology—Washington (State)—Olympic National Park. 4. Natural history—Washington (State)—Olympic National Park. I. Franklin, Jerry F. II. Title.
QH105.W2K57 1992
574.5′2642′0979794—dc20 92-7883

The paper used in this publication meets the minimum requirements of the American National Standard for Information Sciences—Permanence of Paper for Printed Library Materials. ANSI Z39.48-1984.

Olympic Rain Forest was produced for the University of Washington Press by Perpetua Press, Los Angeles.

Edited by Letitia O’Connor
Designed by Dana Levy
Typeset by Wilsted & Taylor, Oakland
Printed in Hong Kong by Great Wall Printing Co. Ltd.

COVER
The Olympic rain forest is a wet realm dominated by trees but also supporting myriad other interrelated life-forms.

PAGE 1
Rain forest trees reach heights of 300 feet; their ages range to 400 years for spruce, 1,000 years for cedar.

Big-leaf maples stand in groves scattered along the valley bottoms.

Contents

Dedication

For Pansy Hudson, Quileute matriarch at the mouth of the Hoh, whose earliest memory is of traveling in her father's cedar dugout canoe;

for Minnie Peterson, homestead daughter, who cooked at peninsula construction camps, packed surveyors and tourists into the inner Olympics, and lived to a cheerful old age on an upper Hoh stump ranch;

for Glenn Gallison, dedicated member of the National Park Service, who loved this forest and worked to protect its future;

and for all the students and scientists who are increasing—and will continue to increase—human understanding of the ecosystem.

Foreword

by Ivan Doig

I AM ONE OF TRILLIONS who live in the Olympic Peninsula's neighborhood. From my habitat east across Puget Sound, I socialize whenever I can with the manifold life-forms of that wondrous community of mountain, coast, valley, river. "Going to the peninsula" has furnished many of the best moments of memory. Hiking in from Lake Quinault to the Enchanted Valley chalet through miles of mid-June snow and watching out the chalet window with respect and, yes, enchantment as three bear cubs and their no-nonsense mother whiled away a couple of hours. A tented night among the marmot whistlers atop Elk Mountain and a triumphant lunch atop rhododendron-boutonniered Mount Townsend. The Hoh rain forest dropping diamonds of dew; Cape Alava with its archaeological pouch of the Makah tribal past; Cape Flattery's windy precipice where you step carefully not to be puffed off the continent. Memories of the Olympic Peninsula spontaneously throng in the mind. *The Olympic Rain Forest: An Ecological Web* exquisitely reveals the nature of the place, which the authors explain "is not only more complex than we imagine, it is more complex than we *can* imagine."

Linkages, interconnections, verily a web of vitality from forest-floor fungi upward through the populous canopy of branches to the fog-spearing tips of giant Douglas fir, hemlock, Sitka spruce, and red cedar—this is the mind-opening exploration offered to us by writer Ruth Kirk and forester Jerry Franklin in these pages. They have both long been included in my gallery of heroes, Ruth for her clear-eyed

books on the lands of America from Death Valley to Glacier Bay, Jerry for his scientific long view into our characteristically hurry-up national use of timber. Their combined vision here shows us intriguing details of the Olympic Peninsula we only thought we knew:

Glacier serapes. "Sixty major glaciers still blanket the high Olympics, an extraordinary amount of ice at this latitude. . . ."

Scuzz. "(Douglas fir) needles are the domain of recently discovered—and nicknamed—'scuzz,' an entire mini-lilliputian realm of microepiphytes and microfauna. . . ."

Ocean-crossing logs. "Logs that escape shore currents float to the open ocean and may eventually enter the great clockwise gyre of northern Pacific currents. . . . Hawaiian chiefs particularly prized Douglas fir logs and used them for the huge outrigger canoes that so impressed European and American explorers and missionaries."

Yet this duo of observers never fails to see the forest for the trees. We are shown time and again how vital the whole community of forest life-forms is—link, link, link, all of them essential: spawned-out salmon carcasses fueling "the nutrient base of the rivers through decomposition, predation, and scavenging"; fallen trees transforming into nurse logs, which over centuries will nurture tufts of seedlings into colonnades of mature trees again; elk browsing the rain forest into balance for the past three thousand years; those minuscule organisms—fungi and their ilk—colonizing every one of the welcoming sixty million needles in the crown of a single Douglas fir, "a billion billion microsites in an acre of forest." Every one of these connections and an infinity more, needed to form this "superb ecosystem found nowhere else in the world."

We are luckier than we deserve in still having an Olympic rain forest in our Pacific Northwest neighborhood. If we came upon this grandly green button of a world spinning through the canyons of space, we would be astounded and exhilarated at such a bonus wonder of the universe. But proximity has bred inattentiveness. As Ruth Kirk and Jerry Franklin quietly put it, "For a century people have cut and replanted the astonishing trees of the Northwest without adequately understanding the entire ecosystem." With this temperate but incisive book at hand, no longer do we have any excuse to be the human equivalent of an ice age disrupting this infinitely intricate world of the Olympic rain forest. Now we know.

Preface

As coauthors we first talked about writing this book in the 1980s. Sometimes we planned to include the entire Sitka spruce forest, which stretches for two thousand miles along the coast from northern California to southeast Alaska, sometimes to limit our focus to the wet, west side of the Olympic Peninsula. In the end we decided on this more restricted coverage.

The broad, flat valleys of the peninsula hold the consummate best of the entire Pacific Northwest rain forest—although this is true partly because so much of the rest has been cut, whereas protection within a national park has kept conditions in the valleys reasonably close to primeval. The valleys remain a sizable, intact patch of a formerly vast ecosystem and are an unexcelled representative of that greater ecological whole. Limiting ourselves to the one area also allows us to sketch various intricacies more completely than otherwise would be possible.

Plant species, wildlife, climate, and stage of forest development vary from place to place and time to time. Diversity is a key word; so is linkage; and so, increasingly, is human attitude. In an earlier age humans lived relatively simply and saw gods in every rock and log and pond. Then technology became more complex and seeing gods more cerebral. Our population grew exponentially, and we came to think of land in terms of economic potential rather than as a matrix formed of animate and inanimate, with us just one part of the mix.

Seers, of course, warned against this. The world is "more beautiful than it is useful," Henry David Thoreau proclaimed over a cen-

Overleaf left
Late winter snow dusts a valley sidewall.

Overleaf right
Lush understory and ground-cover plants and abundant flowing water characterize the Olympic rain forest.

tury ago, and in the 1940s Aldo Leopold added that if we would be successful tinkerers our first rule must be to not throw away the parts, not any of them.

Today the need is to balance human sustenance with the well-being of fellow life-forms and the earth we all share. To do this we need to regain the essence of what we felt when we believed in the animistic gods. Scientific understanding rekindles that awe. It explains much and points toward even more. It gives insight into normalcy and how a healthy land maintains itself. It enhances aesthetic appreciation. And it offers lessons for optimizing productivity in forest plantations while holding onto remnants of natural forest. This is a critical matter: we really need to know more in order to take proper account of the forest landscape, both its use and its perpetuation. Unless such knowledge is enhanced and heeded, we will be in for some very unpleasant surprises.

A forested land has been Jerry Franklin's home since childhood in the Crown Zellerbach pulp mill town of Camas, on the Washington side of the Columbia River. Family vacations meant traveling by 1937 Plymouth to camp within the green serenity of Gifford Pinchot National Forest. His early memories include listening to wind move through Douglas fir and lying on the ground staring up the great boles of trees into the canopy. By age nine he decided to join the Forest Service for a career. Initial realization of that determination came at a time when natural, ancient forests had come to be regarded as unproductive in comparison to younger stands and therefore unworthy of the space they occupied. Management practices favored replacing old growth with even-aged stands of species desirable for timber and managed for board feet.

By the late 1960s, however, glimmerings of professional and public ecological awareness began to offset this concept, and research money became available for studying conifer forests as ecosystems, including old-growth systems. In the last quarter-century, thousands of scientific (and political) reports have accumulated, and thousands of additional acres of wild forest have been clear-cut. Meanwhile, human emotions and convictions grow ever more heightened—and polarized.

Ruth Kirk began life in Los Angeles during the halcyon days when air pollution derived from smudge pots protecting outlying orange groves against occasional winter freezes, not from the exhaust pipes of too many automobiles. Oak-covered hills still characterized the suburbs; there were scorpions and trapdoor spiders to dig from the ground and spring wildflowers to pick and press. The family made trips into the desert in a late-1920s Franklin, an ideal car for the arid Southwest because of its radiatorless, waterless air-cooled engine. Marriage to a National Park Service ranger continued her exposure to western natural heritage, including that of the Olympic Peninsula. Writing brought incentive to probe deeply and seek out experts.

Thus we two came together. Our goal has been to present the forest not as an issue but as a remarkable whole. In this, the compiling and honing of pages has been immeasurably enriched by the studies and suggestions of colleagues and friends, many of whom read sections of the manuscript. We thank in particular Dr. James Trappe and Tom Odell, mycologists with the U.S. Forest Service's Forestry Sciences Laboratory in Corvallis, Oregon; Drs. George Carroll of the University of Oregon and Fred Rhoades of Western Washington University, biologists specializing in studies of the forest canopy; Dr. Mark Harmon, forest sciences professor at Oregon State University; Larry Jones, research biologist at the Forestry Sciences Laboratory in Olympia, Washington; Robert Mowrey, wildlife biologist recently part of a team studying flying squirrels, formerly with the Forest Service in Fairbanks, Alaska, and now with the U.S. Army Corps of Engineers in Seattle; Jeff Cederholm, fisheries researcher for the Washington Department of Natural Resources; Drs. Charles Raymond and Minze Stuiver, glaciologists engaged in Quaternary research at the University of Washington; Dr. Robert Carson, geologist at Whitman College in Walla Walla, Washington; Hank Warren, chief naturalist at Olympic National Park, and park research biologists Doug Houston, Ed Schreiner, Bruce Moorhead, and John Meyer. To them all go our thanks. We also thank Gretchen Bracher, who made the drawings, and various agencies and individuals for generously providing supplemental photographs. Our use of plant common names follows Hitchcock and Cronquist, *Flora of the Pacific Northwest*, and Kruckeberg, *Gardening with Native Plants of the Pacific Northwest*. Scientific names are given in the index.

To the rain forest itself goes our profound respect, along with our awareness of a unique privilege at having attempted to act as its spokesmen. As naturalist Donald Ross Peattie once observed: "Olympic forests are what you imagined virgin forests were when you were a child."

Ruth Kirk—Olympia, Washington

Jerry Franklin—Issaquah, Washington

THE OLYMPIC RAIN FOREST *An Ecological Web*

Yellowing bracken fern and vine maple signal autumn shutdown for perennials and deciduous trees and shrubs.

NEAH BAY
VICTORIA
PORT ANGELES
FORKS
LA PUSH
BOGACHIEL VALLEY
HOH VALLEY
QUEETS VALLEY
KALALOCH
QUINAULT VALLEY
OLYMPIA
GRAY'S
HARBOR

Airborne radar produced this image of the Olympic Peninsula. Its clarity, greater than that of satellite imagery, is a product of radar pulses, which reflect well regardless of weather conditions or darkness.

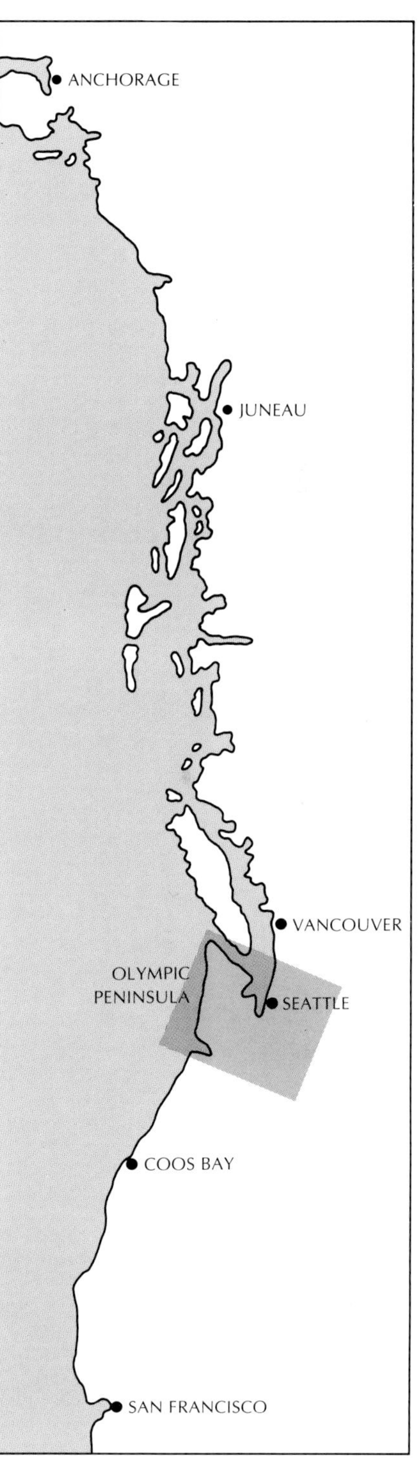

I

A GREEN AND PADDED REALM

Recorded in a Hoh Valley hikers' register:
"Packs too heavy, feet hurt, boots leak, trail too long. We'll be back!"

To step into the rain forest on the west side of Washington's Olympic Peninsula is to enter a green realm where mosses and ferns pad overhead branches and seedlings blanket prostrate logs. Minute soil fungi nourish trees that tower a hundred feet before branching; elk create a characteristic openness in the landscape by their browsing and trampling.

Scientists view old-growth forest as a combination library, laboratory, and warehouse of genetic fine-tuning. Poets liken the luminous light and soaring space to that of a cathedral. Composers could easily score the percussion of rain striking maple leaves and the still surfaces of ponds, but even with an orchestra and a musical form as intricate as a fugue they could scarcely convey the forest's infinite linkages.

Conifers belong to an ancient group of plants that lost preeminence as more adaptable flowering plants developed. In the mild, moist climate of the Pacific Northwest, however, coniferous trees still overwhelmingly outnumber and outsize broadleaves. A forest dominated by Sitka spruce and western hemlock stretches 1,200 miles along the Pacific coast between latitudes 43° and 61° N.

Huge Trees

Hoh, Queets, Quinault, and Bogachiel: these valleys of the Olympic Peninsula cradle what many consider to be the best remaining example of the Northwest's rain forest, a superb ecosystem found nowhere else in the world. This forest reaches its fullest potential on the wet west side of the peninsula where the productivity rate

All members of a natural community plus their physical environment comprise an ecosystem. Left, candy flower, one of the first flowers to bloom in spring; right top, tree frog, a species with toes so sticky they can climb glass; bottom, oxalis leaves fold to reduce exposure to summer heat; opposite top, foam flower and moss carpet a fallen log; opposite bottom, mosses and club mosses festoon vine maple branches.

per unit of area exceeds that found anywhere else on earth. Shrubs, herbs, and grasses, ferns, liverworts, and mosses fill the space beneath trees. The green conquest spreads even to the stony, gray, unstable river bars and to plants themselves, which host other plants called *epiphytes*, Greek for "on plants." Big-leaf maple branches are shaggy with epiphytes, and crevices in the bark of a single living Douglas fir may host hundreds of seedlings, ranging from hemlocks to huckleberries, spring-beauty to foam flower. Alder trunks appear white not because of bark pigment but because of lichens incorporated into the bark. Lichens are a combination of fungi and algae that live as a single organism; the fungi furnish bodily support and the algae produce carbohydrates, needed for energy. Alders expand the fundamental duo into a trio.

The Olympic Peninsula lies an hour's ride by ferry, highway, and bridge west of Seattle. It forms the northwest jumping-off point of the conterminous United States, a sixty-by-ninety-mile thumb of icy mountains, forested lowlands, and still-wild beaches. Drive west from Port Angeles or north from Aberdeen-Hoquiam heading for the rain forest in any month except July or August, and chances are that oncoming traffic will be light. Industrialization and human population remain subordinate to the land itself. Logging trucks still shuttle from forest to storage sites and mills but now carry matchsticks compared with the single-log loads they hauled in the 1970s. Those erstwhile forest giants challenged homesteaders clearing land for farms and sup-

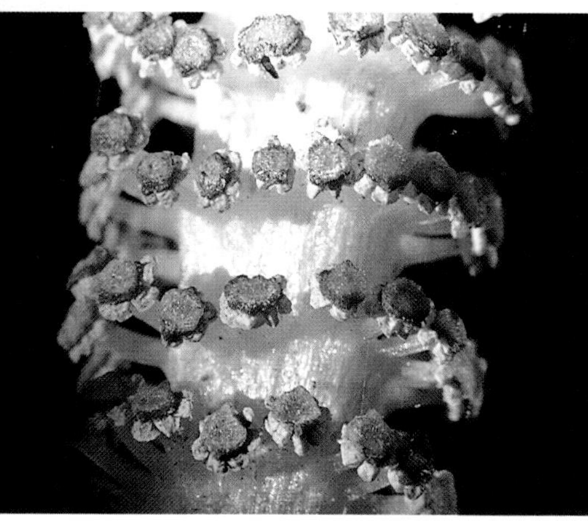

Habitats for plants range from streambanks and the forest floor to the trunks and branches of trees. Top left, money flower grows in wet soil; top right, licorice fern is a common epiphyte; bottom left, close-up of horsetail spore stalk; bottom right, rain droplets coalesce on oxalis leaves.

Opposite
Spring leaves mist alder trees with green.

Mount Olympus (7,965 feet) thrusts above forestlands formerly unbroken except for river corridors. That pattern is still true only in the national park.

Opposite top
Spruce and hemlock dominate valley floors, edged along the riverbanks by red alders and scattered northern black cottonwoods.

Opposite bottom
Alder trunks host lichens, which make them look white; microscopic organisms on the trees' roots enrich the soil.

plied wages for three generations of loggers. Today most of the huge trees are gone.

Large unbroken expanses of lowland forest remain only in Olympic National Park. Although scattered blocks of lowland old growth totalling a quarter million acres have recently been set aside for preservation in the national forest, only one tract is larger than ten thousand acres; a few uncut river corridors and preserves remain on state and corporate land. The forest has become a patchwork. Flying over the peninsula reveals the boundaries of jurisdictions and ownerships, clear-cuts and regrowth, plus a lattice of haulroads and highways. Little but mountains and the upper parts of drainages remain intact, protected within the park.

There, downvalley from the Hoh visitor center, grows a Sitka spruce over thirteen feet in diameter, so huge in girth that encircling it takes eight men standing with arms outstretched and fingertips touching. An even bigger spruce, almost fifteen feet in diameter, grows at the Queets River campground, and another, almost eighteen feet in diameter, stands along the south shore of Lake Quinault in the national forest. The world's second largest Douglas fir (and largest in the United States), fourteen feet in diameter as measured by modern laser technology, grows in the Queets Valley. The largest known western hemlock, almost nine feet in diameter, is in the Quinault Valley.

Red cedars reach huge diameters. Washington's largest tree of any species is a giant cedar—which is hollow—above the north shore of

Sitka spruce ordinarily lives only along the coast, but valleys on the west side of the Olympic Peninsula receive enough rain and ocean-born fog to provide suitable habitat inland. Water vapor readily condenses from the air onto conifer needles, adding appreciably to local precipitation.

Lake Quinault: nineteen feet six inches in diameter and 174 feet high. Inside it resembles a wooden cave: aprons of xylem (hard woody tissue) drape the walls, and the roots of nursling hemlocks and huckleberries hang like stalactites. Crossing the hollow takes nine long steps. Circling the outside of the tree takes nearly forty steps. Such hollow cedars served early-day settlers and travelers as temporary cabins, and in the Elwha Valley (near Port Angeles) one functioned for years as an official U. S. post office. A cedar nearly as large as the Lake Quinault tree grows in the ocean fringe of the park north of Kalaloch, and another stands in grotesque isolation in a replanted clearcut outside the park.

Olympic rain forest trees not only grow huge in girth, they also stand exceedingly tall: to 305 feet for a Queets Valley spruce; 311 feet for a Douglas fir in the same valley; typically 200 feet for hemlock. These heights almost match California's redwoods, considered the world's tallest species because of a specimen measured at 368 feet, and they are close to the height of Australian eucalpti with a documented 312–foot tree. All such acclaim, however, shifts from time to time as additional techniques of measurement are developed and as other astonishing trees are found—or fall.

Despite huge size, however, Olympic Peninsula lowland trees are not particularly long lived. Instead they follow a live-fast, die-young pace that seems long only because it exists on a time scale so different from human life-span. Ages here typically range up to about four hundred years for Sitka spruce and western hemlock, seven hundred for

Spruce (at left) and cedars are representative old-growth trees. Cedars are particularly long lived because their exceptionally broad bases resist toppling by wind. Spruce and western hemlock are more vulnerable because they are shallow rooted and easily thrown; they can also be snapped by wind.

Douglas fir, and a thousand for red cedar. None of these ages is a record for its species. Spruce almost twice this age flourish in southeast Alaska, where soils are less optimum and winters are colder than on the Olympic Peninsula. These conditions cause Alaskan spruce to grow relatively slowly, which seems to foster longevity. Similarly, hemlocks in Alaska live about 50 percent longer than their peninsula counterparts, as do Douglas firs on mid-elevation slopes at Mount Rainier.

Other species throughout the world live far longer than any of these. The championship age among trees belongs to bristle-cone pines in the high mountains of Nevada and eastern California, which are as old as 4,800 years. California's sequoias rate second place with individual monarchs more than 3,500 years old; and alcerces (a type of cedar in the Chilean rain forest) are runners-up that live 3,300 years.

Four Valleys

Magnificent trees comprise the first impression of the Olympic rain forest for most of us, coupled—if lucky—with glimpses of elk. Also immediately noticeable in the Hoh, Queets, and Quinault valleys—and to a lesser extent in the Bogachiel—are broad bottomlands and rivers that shift channels when obstructions divert their flow or when rising water causes flooding. Prudent hikers know about the rising water: to wade across a rain forest river is to risk not returning the same day if a sudden downpour occurs or if high-country sunshine triggers rapid snowmelt. Sculpting by glaciers left

Olympic rain forest valleys are characteristically short for a complete flow from glaciers and snowfields to the ocean. The Hoh reaches only about forty miles from its mountain headwall to the Pacific.

the valleys a mile wide with steep walls that rise to four thousand feet. Sediments gouged and plucked by the ice and washed from mountains and sidewalls form deep fills, now cut into successive terraces by the rivers.

Ice has repeatedly whitened most of the northern continent during the last one to two million years, although details probably can never be known. Glaciation waxes and wanes, and each new advance nearly obliterates evidence of its predecessors. It is clear, however, that an ice sheet pushing out of British Columbia has shrouded much of northern Washington several times, most recently about fourteen thousand years ago. That ice pushed against the Olympic Mountains and split into two vast lobes. One lobe thrust along the trough of Puget Sound. The other followed the Strait of Juan de Fuca, reaching westward to about Neah Bay. At the time, worldwide glaciation was holding so much water on land as ice that the sea level stood three to four hundred feet lower than at present. This exposed the continental shelf, widening the lowland between mountains and surf. At Cape Flattery land extended almost ten miles farther west than is now the case, and the coastline in the Hoh to Quinault region at the base of the peninsula lay thirty miles west of its present position. Earlier ice from Canada had been even more extensive. It stretched along the outer coast as a vast sheet from about Cape Beale on Vancouver Island to Cape Johnson, north of La Push on the Olympic Peninsula.

Local glaciers also have repeatedly whitened peninsula lowlands.

The broad, flat-bottomed cross-section of the Hoh Valley and its steep, evenly truncated side ridges testify to sculpting by glacier ice.

When snowfall in the mountains exceeded summer melt and compacted into ice, glaciers spilled from the peaks down the valleys and onto the lowlands. During the last major glaciation this ice scarcely advanced beyond valley mouths, but earlier alpine glaciers in the Hoh drainage reached the coast and left gravel and silt on top of what are today offshore islands. At the same time glacier tongues in the Queets and Quinault drainages merged with ice from the Humptulips (next drainage south of the Quinault) and formed a huge lobe that spread nearly to the coast. Sixty major glaciers still blanket the high Olympics, an extraordinary amount of ice at this latitude since even the tallest peak on the peninsula is less than eight thousand feet high.

In the Hoh, Queets, and Quinault rain forest valleys the sidewalls are steep; on these drier slopes Douglas fir and Pacific silver fir form the main forest canopy, codominant with western hemlock. At the base of the sidewalls, the oldest valley bottom terraces support what are considered typical rain forest trees: Sitka spruce with western hemlock. Seedlings of Douglas fir germinate among them on patches of raw, disturbed soil, but persist for only one generation, unable to reproduce in their own shade. In contrast, spruce and hemlock are shade tolerant. They form the climax stage of the valleys' forest, the self-perpetuating plant community that will change only if the environment is changed.

Extensive glacial sculpting in the three valleys apparently contributes to growing conditions slightly different from those in the less-

During the Pleistocene Ice Age a glacier inching out of Canada isolated the peninsula to the north and east, apparently preventing a dozen animal species such as grizzly bears, foxes, and mountain goats from reaching the Olympics. Several missing species have been introduced in this century, however, brought in by people.

ABOVE
The ice of Blue Glacier at the head of Hoh Valley is about a thousand feet thick, a product of exceptionally heavy snowfall.

OPPOSITE TOP
A plane flies in supplies to a research station on the Blue Glacier.

OPPOSITE BOTTOM
From the glacier's upper reaches, ridges and valleys appear to be corrugations stretching to the coast. The mountains are formed by the collision of the Juan de Fuca oceanic plate against the continent.

Abundant salal (left) and bunchberry dogwood (right) accompany Douglas fir in the Bogachiel Valley, an indication that conditions there are somewhat drier than in other rain forest valleys.

glaciated Bogachiel Valley. Ice in the three valleys left gradients of only about one percent in their central reaches; bottomlands in the Bogachiel are slightly more sloped and noticeably more narrow. Partly as a result of this legacy the Bogachiel forest differs somewhat from the self-perpetuating spruce/hemlock forest of the Hoh, Queets, and Quinault. Its mossy upholstery and abundant logs shaggy with seedlings (aptly called "nurse logs") are hallmarks of all Northwest rain forest. But the slightly drier Bogachiel forest has Douglas fir rather than Sitka spruce as the codominant with hemlock. Here, Pacific silver fir is not restricted to the sidewalls but is conspicuous also on the bottomlands; salal is a frequent ground-cover plant; and vine maple tends to stand straight and treelike instead of bending into classic rain forest arches and tangles. The Bogachiel forms a transition between the other three valleys and the northern peninsula, a difference that gives it special value.

Geographic orientation contributes to the microclimate of the valleys. The Bogachiel faces storms moving in from the ocean at a slight angle, whereas the other valleys open directly to the prevailing southwesterly winds. Rainfall has not been recorded in the Bogachiel, but it is less than in the Hoh, Queets, and Quinault, where annual precipitation averages 120 to 140 inches and has been measured as high as 180 inches. This is ten to fifteen feet, two billion gallons of water for each square mile of surface. Drizzle and moderate rains are the norm. They fill depressions with puddles, creating habitat for aquatic

Opposite
Moisture is constantly visible throughout the forest, either rising or falling, as here in the Quinault Valley.

Top
Winter sunshine spotlights a young hemlock growing on a snag.

Bottom
A red-breasted sapsucker fluffs its feathers for protection from the cold.

Opposite
Overall, moderation keynotes valley floors.

species from invertebrates to frogs and newts. Runoff from heavier rain causes rivers to undercut their banks and triggers steelhead and salmon waiting at the coast to start their spawning runs.

Precipitation increases steadily with distance up all four of the valleys. At the coast, Kalaloch averages 90 inches a year and Forks, a few miles inland, averages 77 inches. Farther inland, the figures are 142 inches for the Hoh visitor center and 200 inches for Mount Olympus, the ultimate headwall of the valley at 7,962 feet in elevation. Ninety percent of this precipitation falls between September and May. Summer produces only 14 inches of rain, about what people in Los Angeles hope to receive in a year. Moisture in the forest almost always seems to be either coming down or rising back into the atmosphere. Vapor streams along side ridges like tatters of soft gauze and weaves in and out of treetops, and on days that otherwise would be sunny it often funnels in from the ocean as fog and condenses on the forest's billions of needles and leaf tips, then falls with a gentle, almost imperceptible sound. Such "rain-making" by trees is widespread; indeed, in places researchers have found that trees capture twice as much water from fog as falls in rain and snow. Measurements in the Hoh Valley indicate that fog condensation annually contributes about one fifth of the total moisture: 30 inches. That wet contribution of fog alone matches Seattle's total annual rainfall.

Twice as many days each year are overcast in the Olympic rain forest as are clear. Botanists believe this enhances plant growth be-

cause clouds act as a blanket, holding temperatures warmer in winter and cooler in summer. Thermometers have registered as high as 104 degrees Fahrenheit and as low as one degree. But these are extremes. Residents find days that even briefly top eighty degrees worthy of comment, and sustained temperatures much below freezing become equally irksome. Ice often skims winter puddles but rarely becomes thick or lasts long; snow seldom lies deeper than a foot or two, or remains on the ground for as long as a week. Typical January nighttime minimums are in the high forties; July daytime means hover in the low seventies.

Trees themselves have a moderating effect. They hold temperatures as much as ten degrees cooler by day and warmer by night than in adjacent open country. They also intercept snow, slowing accumulation on the forest floor, and dampen the force of wind. Even the gentlest breeze stirs countless square inches of leaf surface and produces a soft overhead rustle often heard several seconds before the breeze itself can be felt. This dance within the canopy cuts the force of moving air by as much as 80 percent.

Temperate Rain Forest

Most forests in temperate moist regions are broadleaf, but in the Pacific Northwest conifers outnumber broadleaves a thousand to one. Their conical shape maximizes exposure to whatever light penetrates the clouds, an adaptation particularly advantageous in winter when the sun rides low in the sky even at noon. Their huge size allows rain forest giants to store massive amounts of water, which can be crucial during periods of environmental stress when smaller trees must shut down their photosynthesis. A single Douglas fir may hold five thousand gallons of water. Even more advantageous is staying green year round. Conifers replace only 10 to 20 percent of needles each year; individual needles can live as long as fifteen years. Contrast this with broadleaf trees, which are dormant for five months each year and then must regrow their entire leaf factories.

Throughout the world there are perhaps a half-dozen coastal strips with temperate rain forests. In addition to those of the Northwest, these include parts of southern Australia, New Zealand, Chile, and also—according to some ecologists—small patches of the Norwegian, British, and Japanese coasts. A glance at travel brochures portraying any of these forests gives an immediate impression of green exuberance, from delicate jades and emeralds to hues so saturated they appear almost black. The kinds of plants vary from forest to forest, but the greenness is constant.

In total cumulative area, temperate rain forest blankets only about seventy-five million acres of the earth; two-thirds of this acreage lies along the Northwest Coast of North America. Botanists list the salient characteristics of all temperate rain forests as:

- wet, cool, acidic soils
- copious networks of flowing water
- relatively little disturbance by wildfire or insect attack

Vine maples hold leaves horizontally to optimize exposure to light during the growing season, a help in offsetting time lost to winter dormancy. Temperate rain forests throughout the world are primarily a coniferous realm merely accented by deciduous growth. Northwest maples do not grow north of British Columbia; red alders continue to southeast Alaska.

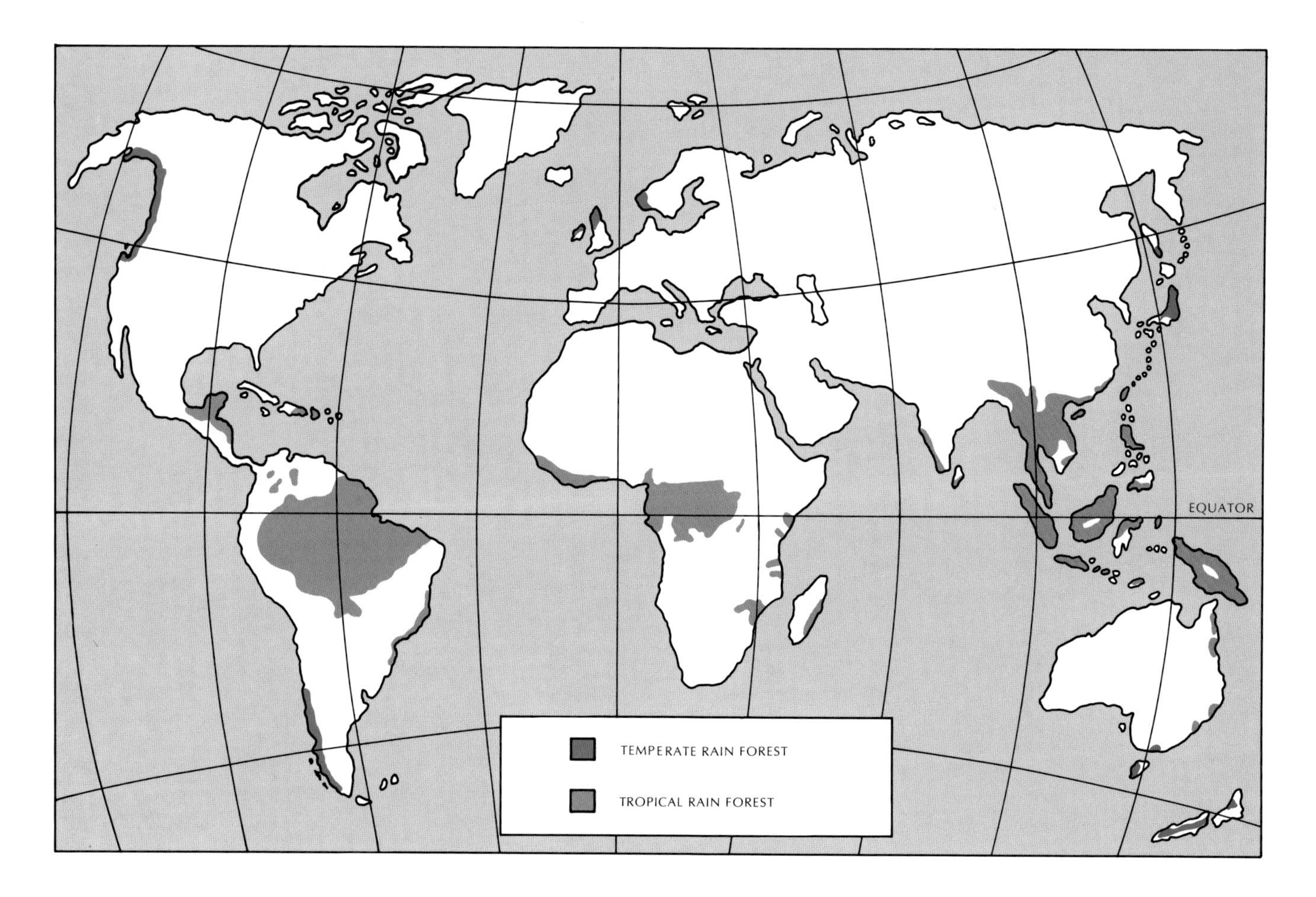

Tropical rain forest vastly outclasses temperate rain forest in extent and perhaps in diversity, but the two forests are now equally vulnerable to changes brought by humans. Only in the last few decades have scientists gained real understanding of either ecosystem.

- primarily needleleaf trees (mainly conifers) rather than broadleaves
- multilayered growth with canopy, understory, shrub zone, and ground cover grading from one into another
- abundant epiphytes and mosses
- large amounts of organic debris covering the ground
- trees that include the largest and longest-lived of their kind

Along a narrow strip from Coos Bay, Oregon (or even the redwood country of northern California) to the Gulf of Alaska, these conditions prevail. Consequently, to know the coastal vegetation of the Olympic Peninsula is also to recognize that of the coasts of British Columbia and southeast Alaska: likeness exceeds difference across twenty degrees of latitude, a greater north-south span of floral sameness than is true anywhere else on earth. Within this sameness, however, certain characteristics set Northwest localities apart from one another.

Southeast Alaska is a bit too cold for optimum forest growth. Also, layers of hardpan only a few feet, or even inches, beneath the surface often block drainage and result in muskegs that stunt tree growth. In British Columbia ice-age glaciers cut deep fiords that have only narrow rims of level land at the base of high, steep walls. Sizeable lowland tracts, optimal for forests, exist only in the lee of Vancouver Island and the Queen Charlotte Islands, topographic positions somewhat protected from storms off the ocean. As in the Bogachiel, the conditions in these areas are somewhat drier than along the outer coast. Their position creates subtle climate differences, so in some places Douglas fir codominates with hemlock, in others either Sitka spruce or red cedar is codominant with hemlock.

Broad lowlands extend the length of the Washington coast, narrowing markedly along the west side of the Olympic Peninsula, where mountains force clouds aloft and release drenching rain. Consequently the ocean side of the peninsula is the wettest part of the state (and of the entire conterminous United States). Oregon's coastal mountains are much lower (typically two to four thousand feet compared with the Olympic Mountains' six- to eight-thousand-foot peaks). Geography makes the Olympic rain forest the crown jewel of its type.

Tropical Rain Forest

TROPICAL RAIN FOREST, which is broadleaf, has a lush quality that outstrips even Northwest rain forest except in terms of living biomass (about 500 tons per acre for Northwest forest compared with 300 tons per acre in the tropics). Tropical rainfall sometimes totals 400 inches per year and deluges the forest at a rate of two inches an hour. Half to three-quarters of this liquid wealth is intercepted by the forest canopy and never reaches the ground; bark typically is thin enough to absorb moisture directly as it trickles down branches and trunks. Treetops harbor birds and monkeys and a legion of other lifeforms ranging from snakes and lizards to bees, wasps, flies, and even mosquito larvae swarming in puddles held high above the ground.

More plant and animal species coexist and interact in tropical rain forest than in any other life community on earth. Indeed, half of the world's known species live there, although the ecosystem covers only 2 percent of the globe. Researchers in Malaysia found 835 species of trees in a single plot two-tenths of a square mile in size, a greater variety than in the entire area of North America. Other botanists counted 42,000 species of plants in just two and a half acres of Panamanian forest.

Tropical climate has been predominantly warm and moist for millions of years, ample time for life to develop myriad forms, specialties, and adaptations, but human use of wood and land now threatens that ancient diversity. Thirty acres of tropical trees fall every minute. "Fighting the green inferno to prepare for the future . . . ," begins a 1970s film titled *Modern Times Come to the Tropics.* The script describes the conversion of Amazon forest into cattle ranches, an attitude that prompted renowned Harvard ecologist E. O. Wilson to caution in a 1991 interview that such action "can be defended (with difficulty) on economic grounds but is like burning a Renaissance painting to cook dinner." Latin America has 57 percent of all remaining tropical rain forest (Brazil alone has one-third of the total). Southeast Asia and the Pacific Islands account for another 25 percent, West Africa for 18 percent.

In appearance, tropical and temperate rain forests share certain characteristics. Their habitats range vertically for two hundred feet or more from ground level to uppermost canopy. Their floors are dimly

Abundant epiphytes characterize both temperate and tropical rain forests. Among Northwest species, big-leaf maples exceed all others as hosts, but researchers cannot yet explain why.

lit. Their trees flare at the base, although for different reasons. In the tropics, trunks send out roots well above ground, producing ten- to twelve-foot stilts. In temperate rain forests, buttressed pedestals form as saplings perched on nurse logs and stumps send their roots inching over the rotting wood to reach mineral soil.

Epiphytes abound in both types of rain forest. In the tropics they include vines, lianas, and lavishly formed members of the orchid and pineapple families. Nobody would mistake tropical epiphytes for those of the temperate zone, but the fundamental growth conditions that let plants live on top of one another hold true in both regions.

Similarly, astonishing arrays of invertebrates dwell in the two rain forests, generally overlooked but vital to forest function. Most of the world's estimated fifty million species of insects belong to the tropics, many of them to its tree canopy alone. Some biologists predict that when the soil and gravel organisms of the Northwest rain forest are fully studied—a process barely begun—the total life-form count here may approach that of the tropical forest.

Protection for the Olympic Rain Forest

"To see the rain forest" is a prime reason given for coming to Olympic National Park, and the park is among the five or six most visited in the nation (along with Great Smoky Mountains, Acadia, Yellowstone, Yosemite, and Grand Canyon, a ranking that changes from time to time). In addition to park status, Olympic is designated as a world biosphere reserve and a world heritage site. Such protection has been far from automatic.

In the early 1900s commercial elk hunting brought the rain forest its first widespread public attention. A federal forest reserve established in 1897 embraced the land in today's national park and national forest and extended westward to the Pacific. The designation came during a period when privatization of the nation's finest timberlands by syndicates and speculators had become exploitive: *They Tried to Cut It All* is the title newspaperman Edwin Van Syckle used for his book depicting logging on the west side of the peninsula. Despite the reserve designation, loggers continued to regard the forest as theirs, and elk hunters continued to send shipments of meat and hides to city markets. Then the hunters, meeting a demand for the peculiar canine teeth of elk—which remain embedded in the upper jaw—started taking just those two teeth and leaving carcasses to rot.

This new waste drew outrage. Hunting already had decimated the elk herds. The roundish, ivory-colored teeth mottled with brown sold for fifteen dollars apiece and were soon dangling from gold watch chains across the male abdomens of urban America. To stop the slaughter, Congress in 1904 debated, but rejected, a bill for Elk National Park; logging interests opposed a park, and their will prevailed. Five years later—and barely before leaving office—Theodore Roosevelt signed an executive order creating Mount Olympus National Monument from the heart of the 1897 reserve. The rest remained as national forest. The new national monument was intended to protect forever the peninsula's spectacular mountains, forests, and herds of elk.

Proponents for changes in the designation started lobbying as soon as Roosevelt's signature had dried. In 1938 the monument was expanded and its name changed to Olympic National Park. However, most proposals—even long after establishment of the park—have urged removing land from federal protection, particularly the lowland forests, which even today are coveted for commodity production. One proposal, in the mid 1960s, would have turned trees from the Bogachiel Valley into sawlogs in exchange for establishing North Cascades National Park. Another, in the mid 1980s, would have caused Hoh Valley skylines inside the park to lose their natural look by cutting timber on adjoining state lands. The view was saved by adjusting ridgetop boundaries.

More than a million acres of old-growth spruce and hemlock carpeted the Olympic Peninsula prior to the arrival of homesteaders just over a century ago. Today only about 3 percent remains, mostly inside the park. Outside the park the U.S. Forest Service has set aside a 1,400-acre research natural area southeast of Lake Quinault. In addition, under the Northwest Forest Plan essentially all of Olympic National Forest's remaining old growth is protected, nearly 260,000 acres. As remnants, however, most of this is in patches and strips, an artificial fragmentation that inevitably turns plants and animals into virtual island dwellers, often with no acceptable way to escape the imposed isolation or to travel from one remnant of habitat to another. Even forest corridors deliberately left to join unlogged acreages may not be suf-

Federal administration of land on the Olympic Peninsula began in 1897 with the establishment of a reserve managed by the Forest Service. Thirty-six years later the National Park Service gained jurisdiction of the inner core of the reserve, which already had been set aside for preservation in its natural state. Olympic Ranger Station, nine miles by trail from the Hoh visitor center, was built at that time as one of three back-country stations. Currently, a ranger lives there in summer and a nearby shelter accommodates hikers.

ficient protection for many plant and animal species which are incapable of traveling along them. Furthermore, any opening into the forest intensifies temperature extremes at its edges, drastically increases wind velocity, and encourages invasion by opportunistic species not typical of intact forest. These effects can be expected to reach two (or more) tree heights into the forest, about five hundred feet. On this basis, nearly half the remaining old growth of Olympic National Forest is "edge affected." Long-term consequences cannot yet be known.

Extensive understanding of undisturbed forest is only beginning to accumulate. Recognizing the need for such information as a yardstick against which to assess human-caused change, the United Nations in 1976 designated Olympic National Park as a biosphere reserve. The action places it among the world's areas acknowledged to be sufficiently intact and representative of their particular biomes to warrant such recognition. These reserves will be used to monitor and study global baseline conditions. Olympic was picked because of its forests.

For a century people have cut and replanted the Northwest's astonishing trees without adequately understanding how the ecosystem's mechanisms and linkages have produced and perpetuated the forest for millennia. The park provides a unique chance to gain understanding of how this system works. Such knowledge carries great

Until recently, its remote location assured the Olympic Peninsula of having a mix of commercial and natural lands, but instead of geography the pattern now depends on human decisions. People must weigh the economic potential for wood products against the value of undisturbed forests for aesthetic and recreational enjoyment and as examples of natural ecosystem composition and function.

potential economic application and reveals details of the remarkable wholeness and interconnectedness of forest life.

No other nation has set aside a resource as valuable as this rain forest for ongoing protection, intending it to be inviolable and treasured not for possible economic return but for itself. That recognition of inherent worth has led to a third designation for Olympic National Park, which in 1982 was dedicated as a world heritage site. This ranks Olympic with more than one hundred such sites including the Galápagos Islands and Mount Cook National Park in New Zealand (also known for rain forest and mountains), as well as Chartres Cathedral, the Taj Mahal, and the Egyptian pyramids—all places designated as unique expressions of nature and culture. This value for man-made and natural heritage was evoked in a Chinese proverb, "People walk on two legs—one nature, the other culture."

II

THE FOREST COMMUNITY

People prefer that which is complicated, growing, and sufficiently unpredictable to be interesting.
—E. O. Wilson, *Biophilia*, 1984

A FOREST IS A BIOLOGICAL COMMUNITY. Trees dominate with their immense size and striking form. But flowers, ferns, and fungi, invertebrates and microbes, mammals and birds, fish and amphibians also constitute a forest. The word *forestis* itself first appeared in A.D. 556, used for a tree-covered area replete with fishing and hunting, activities reserved for the king.

In the Olympic rain forest, wildlife is a readily apparent part of the whole. The jackhammering of woodpeckers (as fast as twenty strokes per second) and the trilling of winter wrens commonly counterpoint the hush. A glance is all it takes to find the browsed ends of sword fern fronds and huckleberry and vine maple twigs, signs of elk and deer. Having only lower incisors, the animals snip vegetation by breaking it off against upper gums. This frays the ends of twigs. Occasionally a tree partly stripped of bark marks where a bear has scraped cambium with its teeth, savoring the sweetness of rising springtime sap; or a mauled young spruce testifies to a bear having straddled it and ridden it to the ground as a living belly scratcher. Hikers who wade the Queets River or the south fork of the Hoh can still expect to find cougar tracks, sometimes superimposed onto boot tracks. Such

The structure of old-growth forest provides varied and constantly changing exposure to light and storms. This affects plant growth and enriches habitat for wildlife.

occurrences more nearly signify forest health than an ominous, stalking prelude to attack. Cougars, like other animals, readily accept the convenience of a well-built trail.

Old-Growth Ecosystems

Old growth differs greatly from a forest managed by humans. Its ecosystem takes two to three hundred years to develop along the Northwest's wet, coastal strip. Ironically, regional economics dictate cutting trees after about forty or fifty years, even though production of top-quality wood takes two or three times that long. Old wood has straighter grain, fewer knots, and greater strength and decay resistance than young wood. Regardless, waiting is uneconomical because it ties up investment for too long.

The commercial forests of recent decades include only species valued for timber and fiber, and they feature even-aged stands that are the product of a neat start-grow-harvest sequence. Ancient forests are markedly different. They lack uniformity; their growth is at all stages; their masses and spaces irregular. Trees of various kinds, sizes, shapes, and ages form an enormous, many-tiered canopy that is not just a green umbrella with leaves but a structure distinctive in its own right and the home of myriad life-forms. Shrubs provide an understory; herbs and grasses, ferns and mosses, a lush carpet. Massive fallen logs and branches litter the forest floor. Standing dead trees (called snags), rotting stumps, and logs contribute actively to present and future, for deaths here are not endings. Huge woody debris acts as a reservoir

The wildlife of the rain forest ranges from mountain beavers burrowing in the forest floor (left) to occasional wood ducks (right top) and great blue herons (bottom), both of which rest and feed at quiet ponds and streams. Old-growth forests support a greater diversity of animal species than young stands or commercially managed forests.

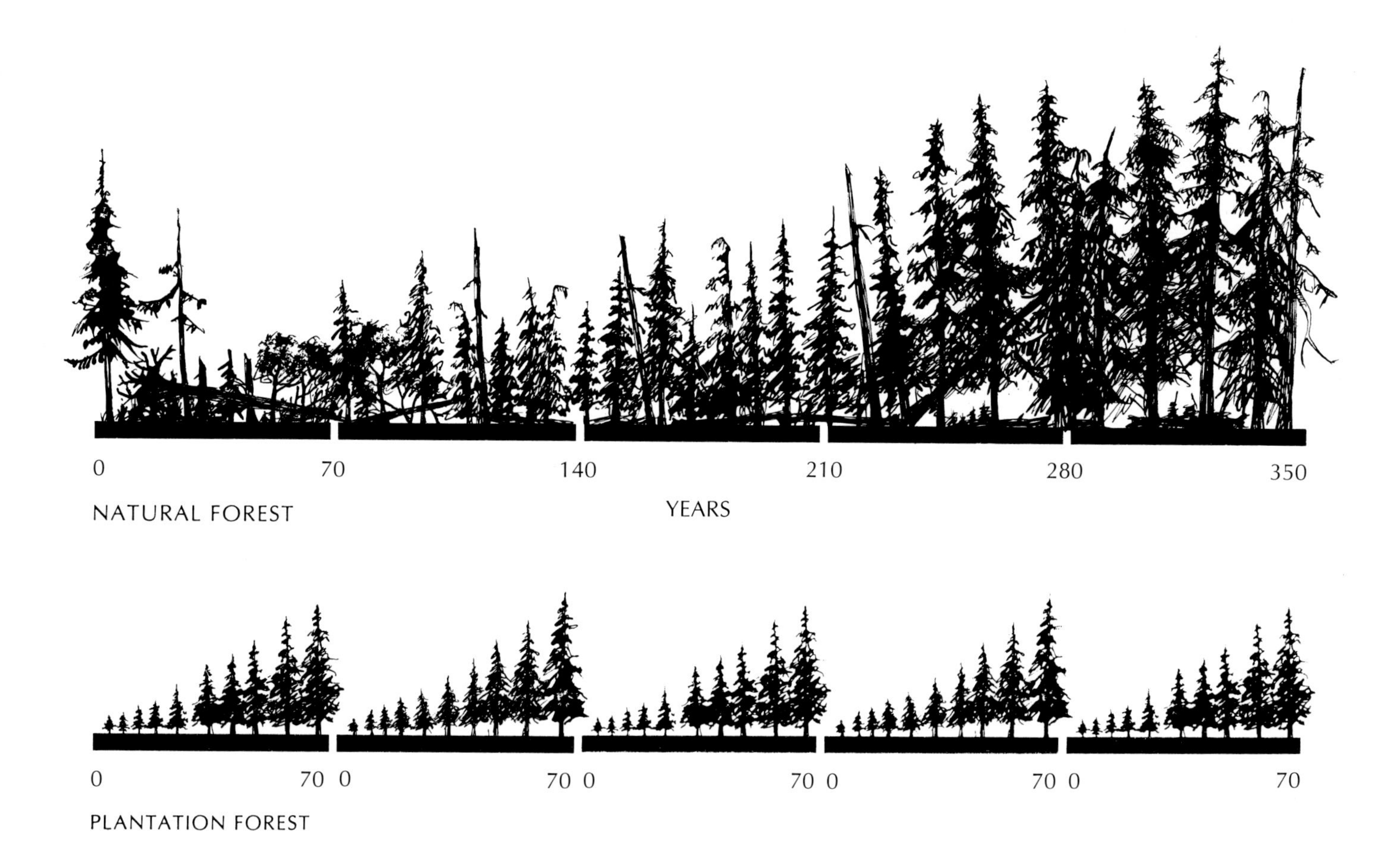

In recent decades plantation forests have been managed for homogeneity and short cutting cycles. Their trees rarely are allowed to reach maturity (about 100 years) or to become old growth (at least 200 years). Natural forests are a mosaic of ages and stages. Even dead trees continue to contribute to the community for centuries.

of stored moisture, energy, and nutrients, which gradually become available for release back into the ecosystem. The debris also is an important part of the forest's architecture: logs and stumps provide barriers, runways, shelters, perches, and dams. Old growth is a self-perpetuating stage in the development of the forest, the culminating and most stable stage of a continuum. Its significance derives less from the age of the trees than from that of the ecosystem. It is the community that is ancient.

Several vertebrate species—particularly birds and bats and probably also amphibians—are strongly linked to old-growth systems. Some seem to live primarily where habitat has reached this stage. Others need old growth at certain times, such as when nesting, or require that environment to produce the large broods necessary for populating adjoining, less desirable ecosystems. Of these species, northern spotted owls currently draw the most attention. Old growth appears to be their prime habitat. Therefore as it decreases, the welfare of this species also decreases.

The birds are moderate sized—about as big as crows—but need large home territories. In western Washington each successfully feeding and nesting pair uses from 2,700 to 4,500 acres; the owls seem unable to adapt quickly to patchwork forest dominated by young growth. Wildlife biologists estimate the state's total spotted owl population at only 500 pairs. Their numbers on the Olympic Peninsula—about 90 pairs—are expected to decrease because of habitat reduc-

Martens, large weasels rarely seen by humans, feed near streams and rest in cavities within large trees. Quick, agile and catholic in taste, this species readily captures prey ranging from squirrels and birds to frogs and beetles. Martens often travel for miles through the canopy without coming to the ground.

tions, a fundamental limitation that even the arrival of recruits from outside the peninsula could not offset (and significant recruitment is unlikely because of the peninsula's geographic isolation).

Impending crisis throughout the owls' range, which is located from northern California to southern British Columbia, prompted extensive research following passage of the federal Endangered Species Act in 1973. By that date refinements in research techniques included fitting birds with tiny radio transmitters to follow their movement and playing their recorded calls (or imitating them) at selected locations in the forest to elicit responses. Reams of field data have resulted; computers are permitting analysis; and conclusions are proliferating. The multi-layered canopy of old growth provides owls with needed maneuvering space and options. Perching in the understory gives maximum shade on hot days; moving into the overstory, close against tree trunks and roofed by overhead branches, offers protection in cold or rainy weather. Feeding occurs at all levels from the ground to the treetops; prey species include occasional spiders and songbirds, but flying squirrels, pack rats, and other small mammals dominate the menu. Surpluses not eaten at one meal are cached on a branch for a future meal. Cavities in the wind-broken tops of large trees and where branches have broken from the trunk provide nest sites; so do platforms of twigs and debris caught on the tops of large branches or built of sticks by other birds or mammals. Individual spotted owls live fifteen years, but since they do not breed until their third year and only

Marbled murrelets feed at sea but hatch their young in the forest. At this latitude they nest exclusively among the luxuriant mosses and lichens of huge, aged trees, so ecologists consider them dependent on old growth.

about 11 percent survive beyond the first year of life, the population continues to decline. Researchers currently place the number of owlets at only a precarious 0.49 per pair of adults.

Inflexible requirements may be dooming spotted owls; and barred owls, which are classified in the same genus as spotted owls, may be speeding this fate. Similarly sized, these birds are not dependent on old growth. They thrive in young forest and also penetrate remaining bastions of mature and old growth. Unfortunately for spotted owls, the barred owls have recently moved onto the Olympic Peninsula and show signs of competing severely, perhaps even to the point of threatening local survival for spotted owls.

Owl studies have focused widespread attention on the matter of radically altered environment. They probe complexities that have been millennia in the making and are actually at the heart of what has sometimes been portrayed as a sudden competition between a small bird and a large industry.

The real issue is not that simple—nor is it limited to spotted owls. Pileated woodpeckers, the largest North American woodpecker after ivory-bills (which are probably extinct), are also strongly linked to old growth. Ecologists consider their presence an indication of old growth west of the Cascade Range. These particular woodpeckers find ideal sites for chipping out nestholes only in large snags, and there they also feast on their favorite food, carpenter ants. The ants, as well as the birds, depend on dead wood. They need it for excavating galleries

Amphibians both absorb and lose moisture through their skin and so are restricted to wet environments. Tailed frogs (above left) fertilize eggs internally; their "tail" lets males transfer sperm into females' bodies. This species seems sensitive to the increased silt and warmer water associated with logging; their future may depend on preserving undisturbed watersheds. Cope's giant salamanders (right) also seem dependent on the pristine pockets of environment where they have evolved. Generally aquatic, they occasionally lose gills and become terrestrial, as shown here.

OPPOSITE
Trees that are dead but still standing, called snags, are crucial for wildlife that need cavities for nests, dens, or resting places. For many species only trunks of large diameter seem to offer sufficient insulation. Some prefer hard snags with bark still relatively intact and branches clinging; others select more decayed snags situated with nearby protective cover in the middle and lower stories of the canopy.

and maintaining "herds" of aphids, which they milk for a sugary secretion. They escort the aphids outside to graze in summer and keep them inside during winter, like cows in a barn. Nor is this aphid-ant-woodpecker chain the end of the linkage: saw-whet owls, bats, martens, and flying squirrels all use abandoned pileated-woodpecker holes as nest sites and roosts, and may compete for them so seriously that the presence or absence of the holes affects the population size of these secondary users.

Vaux's swifts also use tree cavities found in old-growth forest. They are five-inch swallowlike birds that fly through the canopy reportedly as fast as forty miles per hour, gulping flying insects. For nesting they require snags that are tall and hollow and at least two feet in diameter, an assurance of thick walls as insulation.

Marbled murrelets, which are seabirds, nest in old growth, but in living trees, not snags. Small, stubby relatives of puffins and auklets, these birds feed offshore on small fishes and crustaceans, but fly into the forest to nest. As early as 1925 loggers working near Bellingham (north of Seattle) found a murrelet egg in the forest, and over the next several years occasional observers reported the birds carrying fish inland, seemingly to nests. Nobody found a nest or knew for sure where they went. Then in the 1950s a logger in the Queen Charlotte Islands of British Columbia noticed an adult murrelet in the debris of a newly cut, large hemlock, and in the 1970s loggers on Vancouver Island (also in British Columbia) discovered a downy chick high in a Douglas fir.

Since then nests and fledglings have been found around the

northern Pacific Rim, including on the Olympic Peninsula. In places without conifers, nests have also been found in rock cavities and on steep tundra banks, but wherever possible the birds seem to nest only in the huge, ancient conifers typical of humid, virgin forests with flowing streams. They lay their eggs on top of epiphytes close to the trunk, where they are protected by an overhanging branch or by a slanting section of the trunk. Such requirements mean that in this part of their range marbled murrelets need trees at least 150 years old because the thick aerial cushions of mosses and lichens typically take at least that long to develop. And these murrelets rate only such cushions as prime real estate for a family.

Certain forest-floor and stream-dwelling amphibians may also be linked to old growth, although studies are far from complete. Such species include Cope's giant salamander, Olympic salamander, Van Dyke's salamander, and the tailed frog, all of which are considered relict species, remnant populations apparently now isolated from others of their kind. Such creatures become dependent on the pockets of environment that are their entire realms, and in response to these restricted habitats they develop unique characteristics as centuries and millennia go by.

Most Cope's giant salamanders, five to six inches long, live only in water. Few ever progress beyond the larval stage; they reach sexual maturity with gills intact. Olympic salamanders, smaller relatives of Cope's giant salamanders, live mostly in seeps and small streams protected by rocks. Females have the peculiar habit of laying eggs communally, several of them gathering specifically for the occasion. Hatchlings take about four years to lose their gills and transform from larval to adult stage. Van Dyke's salamanders are more terrestrial and follow a different pattern; individual females lay and brood eggs under rocks or in decaying logs. Their hatchlings emerge as miniature, half-inch replicas of adults. They go through the entire larval stage while still within the egg and start life after hatching without gills.

Tailed frogs—brown, voiceless, and less than two inches long—are considered, along with three New Zealand species, to be the world's most primitive, yet most specialized frogs. They are holdovers from Jurassic time, about 180 million years ago. Eggs are fertilized inside the female's body. Tadpoles have large suckerlike mouths that let them cling to rocks in the rushing sidewall streams that are their only habitat. Some take as long as four years to change into adult form; sexual maturity takes about eight years. Individuals may live as many as fourteen years.

All such creatures with specialized needs and adaptations demonstrate that Northwest forests and wildlife have evolved together for a long time. Reduced populations leave some species with little resilience for enduring temporary catastrophe or increased competition. Many appear too finicky to adapt to today's disruptions, or they lack mobility to travel to other, possibly suitable locations. Not only do wildlife species need trees, they need trees at all stages of life and death—in short, the whole interacting old-growth community. If that

Chickarees, also called Douglas squirrels, are most abundant in old-growth forests, possibly because older trees produce more cones for food than younger stands. The squirrels harvest cones singly and use specific perches for stripping them to reach and eat the seeds. Litter from the cones produces conspicuous mounds. Squirrels also cache surpluses underground, in stump cavities, and sometimes even in water, which keeps cones from drying and shedding their seeds.

community is lost, they may be lost. Perhaps, given time, some could adapt to a new set of realities, but people now set the timetable and seem loath to wait. On the Olympic Peninsula the species that apparently are closely tied to old growth and may therefore be increasingly vulnerable includes: northern spotted owl; pileated woodpecker; Vaux's swift; marbled murrelet; goshawk; several bats (silver-haired; hoary; long-eared myotis; long-legged myotis); northern flying squirrel; coast mole; marten; fisher; Cope's giant, Olympic, and Van Dyke's salamanders; tailed frog.

These species are part of the forest's natural diversity. The trend toward their loss is of concern because decreased diversity signals degrading habitat—for us all.

Canopy Life

THE CANOPY OF THE RAIN FOREST presents an array of vertical surfaces, horizontal surfaces, openings, cavities, and tangles. Humans seldom scrutinize its complexity unless kinglets lisping in the treetops or chickaree squirrels bombarding the trail with spruce cones provide incentive to look straight up. Genetic heritage sets the basic parameters for each tree, but individual growth is molded by exact growing conditions, diseases, storms, and competition with nearby trees.

Time compounds the mix and results in a canopy not only structurally diverse but rich with surprisingly intertwined lives. The crown of a single Douglas fir has sixty to seventy million needles. In an acre of forest this means a billion billion tiny surfaces that are colonized by

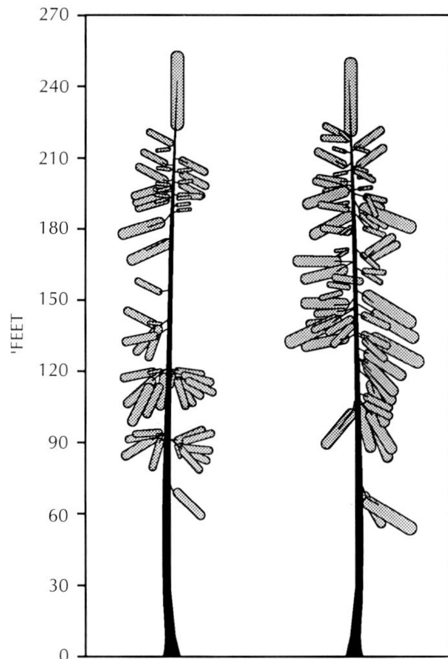

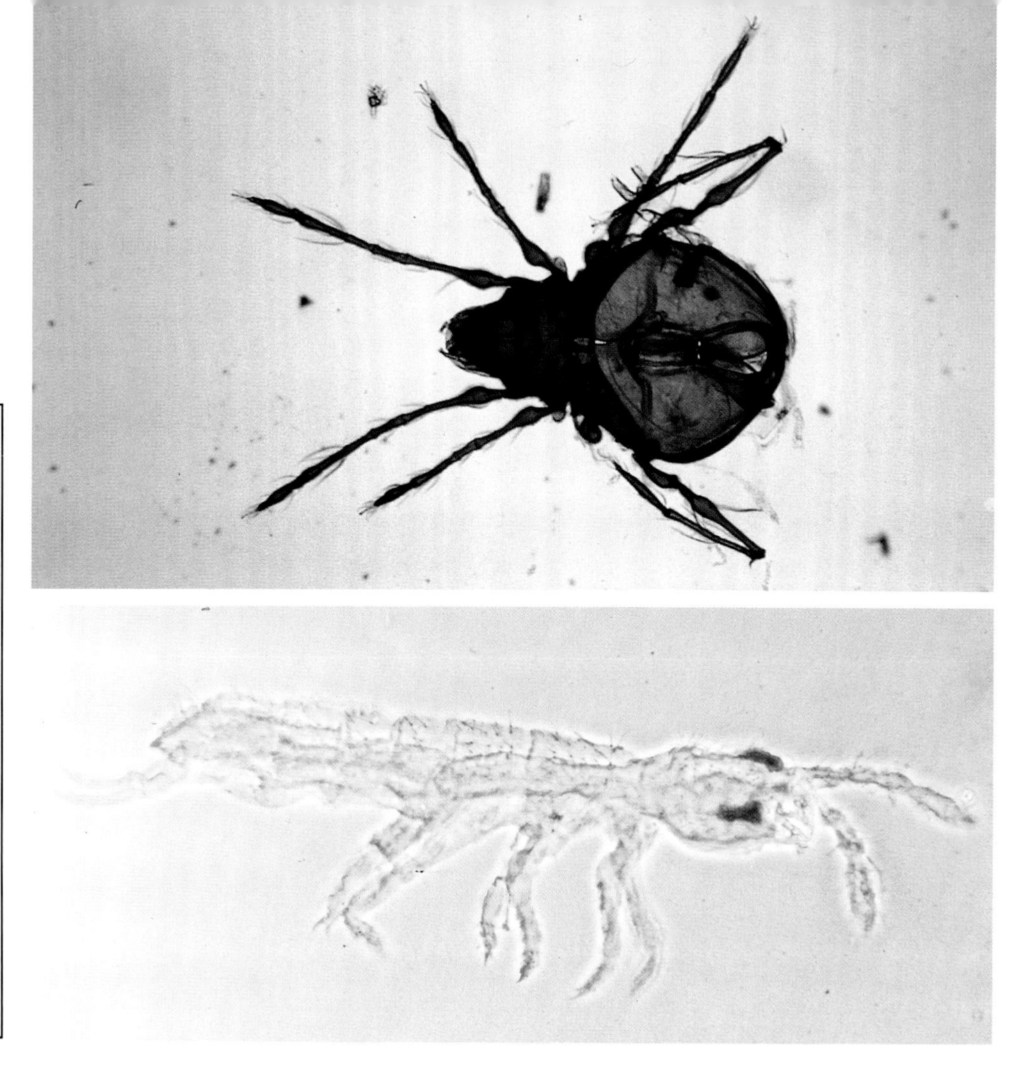

Schematic profiles of an old-growth Douglas fir tree illustrate the wide variety of surface orientations, spaces, and structural overlaps within a single tree. The array of habitats meets the requirements of diverse life-forms and distinguishes old growth from the comparatively uniform character of younger forests and tree plantations. Needles host mites (top right) and springtails (lower right), and secretions from such minute invertebrates apparently nourish "scuzz" fungi (opposite top left). Shown at 600 times actual size, the underside of an eight-year-old Douglas fir needle (opposite top right) reveals an entire web of microorganisms concentrated in surface depressions and plugging stomatal pores. Near the top of the food chain, spiders (opposite bottom) apparently prey on insects drawn to the canopy to feed on scuzz organisms.

minuscule fungi, algae, yeasts, and bacteria. These organisms are called *epiphyllis*, "on leaves." Other fungi live inside the needles as *endophyllis*, "in leaves." The entire domain has been recently discovered and nicknamed "scuzz." It is a mini-Lilliputian realm of microepiphytes and microfauna, a life system complete within itself, yet of consequence beyond itself: wheels within wheels. Functions and interactions are still largely unknown, but at least some scuzz organisms seem to "pay rent" to their landlords in the form of natural insecticides and to enrich the forest as a whole by supplying nutrients. Several scrounge nitrogen from canopy solutions and pass it on. Others absorb it secondhand as it leaches from the cells of their dead neighbors.

Epiphylls and endophylls also form the base of a house-that-Jack-built chain advantageous to conifers. Mites, springtails, and amoebae graze the scuzz; some of them seem to be a valuable food base for spiders and predacious insects. This may be of major importance when leaf-eaters arrive in summer to nibble needles and suck juices. Their invasion might wreak severe damage were it not for the predators waiting almost as a standing army, already well positioned. As a result of this interaction, natural old-growth canopy loses only about one percent of its productivity to insects, a figure markedly below the loss in nearby forests managed by humans (and far, far less than occurs in forests east of the Cascade Range, where the ecosystem is considerably different.) Herbivorous insects, such as hemlock loop-

Lichens camouflage the nest of a rufous hummingbird, the only hummingbird species found in the forest. It feeds on nectar and insects.

Fallen to the forest floor, the common lichen lobaria will be quickly consumed by organisms from microbes to elk. It is rich in nutrients and easily digested.

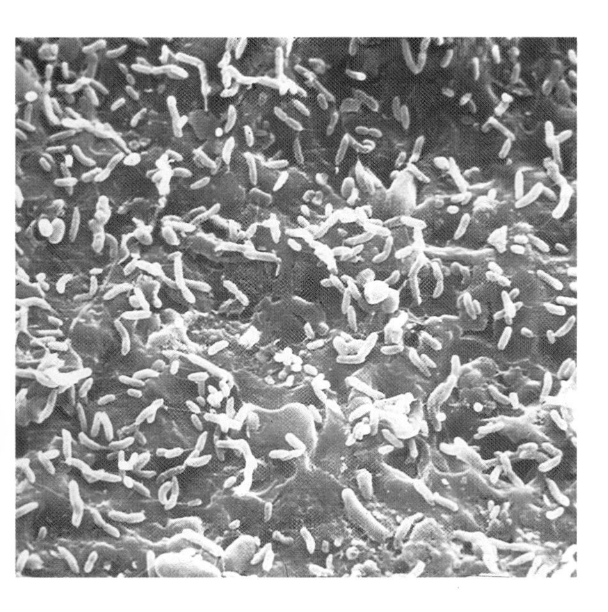

Top
Licorice fern is one of many plant species that live only in the forest canopy, not on the ground.

Bottom
Bacteria crowd the surface of nitrogen-fixing lobaria.

ers, are present in the rain forest, but rarely cause serious damage. The scuzz-predator link probably is only part of the explanation; more research is underway.

Lichens in the canopy also play a major role in the forest community. They are a union of certain fungi and algae—and sometimes also cyanobacteria (the new classification for what formerly were considered blue-green algae). These component organisms seldom live independently. Rather, they merge into a distinctive life-form that may appear crustlike, leaflike, or mosslike. In the rain forest the most common example of the partnership is *Lobaria oregana*, which litters the ground in winter with what look like leathery cabbage leaves. The cumulative mass of this one lichen may be equivalent to one-fifth of its host tree's foliage.

Lobaria's green, upper surface contains its photosynthesizing, algal layer. Its whitish fungus surrounds the algae and has small spherical nodules of the cyanobacterium scattered through its tissue. This third partner is *Nostoc*, an organism able to incorporate atmospheric nitrogen into part of its proteins, which are then directly useable by plants and animals. Only a few kinds of bacteria and cyanobacteria can do this; other life-forms are dependent on them (or on the nitrates now produced industrially). The natural "fertilizer" delivered to the forest floor by falling lobaria amounts to five to nine pounds per acre per year. This lichen also benefits elk and deer; it is tasty, nutritious, easily digestible, dependable, and readily accessible—in short, absolute largess, particularly valuable in winter.

Flecks that break from existing lobaria start anew if they alight in a suitable place, but they grow slowly; reaching cabbage-leaf size may take half a century. Transplanting this lichen to young commercial forests rarely succeeds. In the rain forest a complex of factors that include temperature and moisture seem critical for establishment. Lobaria is active only below air temperatures of 60 degrees Fahrenheit, and humidity must be high enough to hold internal moisture content at 70 percent or more except during intermittent summer drought. Old-growth coniferous forest meets these requirements because great canopy depth assures shady sites, and the mammoth trees garner and hold water (more than an impressive 250,000 gallons per acre), which further cools the air and helps maintain humidity on otherwise dry days. Canopy irregularities create turbulences, which stimulate the collection of rain and fog. Varying branch and trunk angles and tilts present a mosaic of surfaces: water catches on upper sides; lower surfaces are relative deserts.

Big-leaf maple trees outdo all other species as hosts for epiphyte gardens. The physical and chemical qualities of this tree's bark and branches are probably the reason for the compatibility of the host and epiphyte, but nobody really knows. Whatever the factors, big-leaf maples support the lush green cushions and drapes that draw people from all over the world to the Olympic rain forest. Their most showy "moss" is actually the club moss *Selaginella oregana*, a plant closer to ferns than to moss despite its appearance. This plant forms the great-

Big-leaf maple trees support epiphytes that weigh a ton or more when wet. Draperies are mostly the club moss selaginella (top left and right), but also mosses, lichens, and liverworts. Licorice fern (bottom) yellows if summer brings several hot, dry weeks, but it freshens with new fronds as soon as normal wet weather returns.

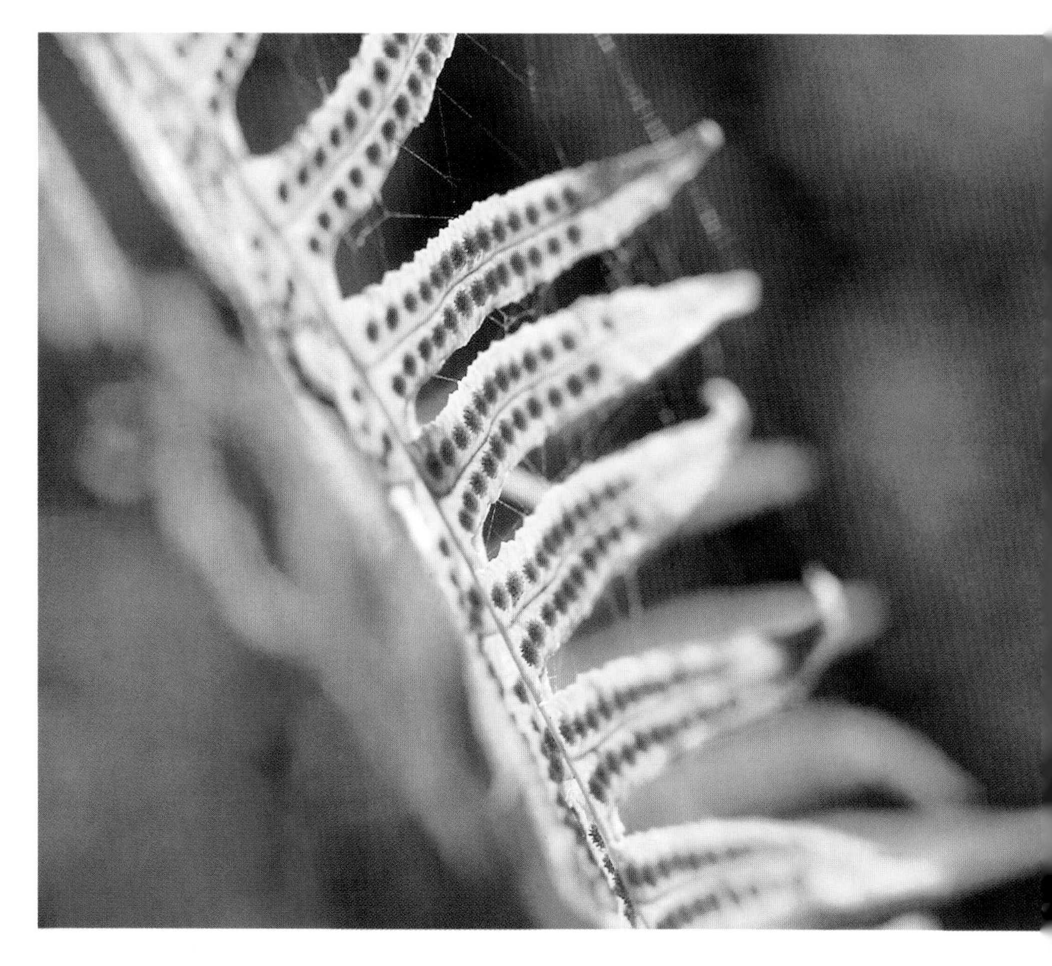

Occasionally ground-cover plants like spring-beauty (above) flourish epiphytically on big-leaf maples. Scientists have identified a hundred kinds of major epiphytes in the Olympic rain forest.

est aerial swags and contributes most of what one visitor described as the "designer look" of this forest.

Surprisingly—despite a long-held assumption that trees act only as passive toeholds—the drapes directly benefit their hosts. This was discovered in the 1980s when Recreational Equipment, Seattle's renowned outdoor equipment cooperative, donated ropes, ascenders, and seat harnesses to Professor Nalini Nadkarni, then a University of Washington graduate student, who used the gear not for mountain climbing but to outdo Tarzan in moving freely through the canopy. In the Hoh rain forest outside the national park, she peeled back mosses, selaginella, ferns, liverworts, and lichens padding branches and trunks—and found maple roots! Whole networks, from tender new root tips to thick woody strands, laced the branches and reached into the epiphytic mats. Individual roots measured as much as twenty feet long. Some angled upward. A few reached from the canopy into the ground.

Expanding her rope web and hauling self and equipment up and down, Nadkarni mapped and measured the surface structure and branch sizes and positions of several maples. She also checked mature alders, cottonwoods, and vine maples. All had roots reaching from their trunks into their mats of epiphytes. Examined under a microscope, all roots showed the usual anatomy of cortex, xylem, phloem, root caps, and root hairs. Why? Apparently for the same reasons governing roots that reach into the ground: to absorb and conduct water and nutrients, abundantly available within the green upholstery.

More than 130 species of epiphytes live in Hoh Valley trees. They lie along branches as pads typically six to ten inches thick. Their total weight often is four times that of their host tree's foliage, a heavy burden especially when wet (which is almost always). The biomass of epiphytes in a maple grove is double that of understory shrubs and ground-cover plants in the same area. Many epiphytes have leaf forms and extensions that catch and hold water, and surface cells that drink in moisture quickly and shut down equally quickly. All thrive on nutrients available in rainwater, dust, and litterfall through the canopy. This was believed the whole story—until Nadkarni left the ground.

Subsequent research indicates that tropical rain forest trees and those in other temperate-zone rain forests also send feeding roots into their epiphyte mats.

Elk and the Understory

Elk are the "landscape gardeners" of the forest; their browsing and trampling shape its openness and character. During winter the forest offers a limited quantity of forage that is only moderately nutritious but sustains the elk population at an appropriate, fairly constant level. This ecological balance is not universal for ungulates. In other places, such as grasslands, herds often go through periodic cycles of overpopulation followed by die-offs. Under such conditions balance is achieved over time by a swinging of the pendulum: favorable years with lush growth are followed by drought; populations build, then crash. In the rain forest, on the other hand,

Alder flats lining riverbanks are an early stage of forest development. Elk heavily browse brush and sword fern there (above left), creating meadows where black bears (right) graze immediately after leaving winter dens. Tender new grass provides the most readily available early-spring growth and seems particularly appealing to bears following their long quasi-hibernation.

Opposite
In most of the rain forest, salmonberry (top) and devil's club (bottom) survive only where beyond reach by elk. Humans find both plants painfully prickly to touch, but their thorns apparently offer little protection from animals. Even slugs climb devil's club stems to devour leaves, which release nutrients and stored energy quickly and have no secondary compounds or wax coating that would reduce digestibility.

forage and herd size hold steady, well balanced with one another.

Pressure from carnivores is also a factor in establishing population equilibrium, but it is a less important control in the rain forest than was formerly assumed (and than may be true elsewhere). Cougars here prey on aged and weak elk and on calves, the classic pattern linking meat-eaters and plant-eaters. Black bears and bobcats, and perhaps also coyotes, probably seek calves but not adults. Grizzly bears seem never to have reached the Olympic Peninsula; there is no historical, ethnographic, or archaeological indication of them. Wolves are gone, locally exterminated by the 1920s, although an occasional sighting is still rumored, and some biologists recommend restoring wolves to the ecosystem.

Humans have been the predator with the greatest effect on elk. In 1905 Washington instituted a bounty for predators, a program intended to protect elk, deer, and homesteaders' livestock. The real protection that elk needed, however, was from guns, and a ban on hunting them went into effect that same year. By then, more than a decade of market hunting for hides, meat, and teeth apparently had already cost the valleys most of their elk herds. The forest understory, no longer heavily browsed, grew rank. But when the hunting stopped, populations rebounded. Some observers thought there might even be too many elk in places. In the 1930s Lena Huelsdonk Fletcher, daughter of Hoh Valley homesteaders, commented on this in a letter to the chief forester's office in Portland. Maybe there had been too much

Vine maple limbs often root when the weight of snow or fallen branches brings them into contact with the forest floor. This creates distinctive arches and tangles that contrast with the generally open character of the rain forest understory.

hunting before, her letter said, but "30 or 40 years ago . . . thousands and thousands of acres were covered knee deep with forage, the very roots of which have now disappeared before the hunger of the herds of game."

Federal biologists also reported that elk herds were more "destructive to the range" than deer, which are much smaller and are outnumbered by elk three to one in this forest. Elk "concentrated too much on favorite plant species"; their trampling was "especially bad" in wet places. Huckleberry suffered most of all because it "never grows above the reach of elk" and therefore might not survive the onslaught. At the time people saw the forest almost exclusively in terms of productivity: land to be cleared for farms, trees to log, and berries and game to harvest and hunt. They urged that resource use should be maximized.

Other biologists looked for the complexities that affect equilibrium in this particular forest. Their work, begun in the 1930s, is continued today by National Park Service biologists and university ecologists. For instance, elk had been supposed to roam in loose, ever-shifting aggregations, whereas in the rain forest valleys they actually form a matriarchal society of half-sisters and young brothers, second cousins and grandmothers. They have lived together from calfhood, learned in detail about the food and shelter of their particular area, and passed local expertise from generation to generation.

Elk cows know where to forage and find shelter, season by season. They help each other with the demands of motherhood (including babysitting) and heed the judgment of a leading dowager who often stays quietly in the center of the group but moves to the edge if she senses danger. She knows where the best sword fern is, which route around a tangle of wind-toppled trees is easiest, how to find banks gentle enough for calves to climb, and where the river can be waded safely. If they must swim, cows sometimes stay on the downstream side of their young as protection, although at other times they seem to abandon their young during the crossing and leave them to manage as best they can. A group's range—at least in the Hoh and Queets valleys—averages about six to twelve square miles and belongs to about fifteen to thirty elk. Cows apparently stay in these socially unified groups, leaving only occasionally to sojourn briefly with a neighboring group. Such absences entail little distance and seldom last more than five or six days. Bulls follow a different pattern, which has not yet been studied in detail. It is evident, however, that most bulls—or perhaps all—leave their natal groups while still young.

Because they are both smaller and fewer than elk, deer affect the forest less. They can, however, reach into nooks inaccessible to elk, an ability significant in winters with sparse forage and during severe blowdowns when fallen trees sometimes trap elk. By lowering their heads and stretching their necks, deer become so streamlined that they can even bound through vine maple without altering stride.

Elk herds live in the rain forest valleys year round except for occasional brief periods in winter. Some groups then move into the relatively dense Douglas fir and hemlock along the base of valley walls facing south, where snow conditions may actually be less troublesome than along the adjoining river bottom. This is true because cold air flowing downslope often settles in the valleys and creates a decided chill, which may prevent snowmelt. If a crust forms, movement through the snow can become exceedingly difficult for heavy, hooved animals. Elk in the upper ends of the valleys, above the elevation of classic rain forest, find their best wintering conditions along the river, then move into the high country as the snow melts and subalpine meadows burst into summer glory.

Antler size seems to be an advertisement of male vigor that is appealing to females. Favorable summer nutrition leads to optimum fall antlers, which signal the prime condition of a bull. The clarion bugling of bulls in rut also stimulates female hormones—and causes lesser bulls contemplating a challenge to back off. Other bulls may withdraw from the sheer decibels, or possibly because they recognize individual voices and are aware of relative standing. Bellowing well requires strength; it is a flaunting that contributes to success in mating rituals. The word *rut* is from the Latin verb *rugire*, "roar." Pawing the ground also figures in the rites of fall, as does wallowing and spraying urine (which may have pheromones, sex-related hormones, that proclaim a bull's condition).

Most of the time mature bulls roam rather independently from both the established groups and from each other, a spacing mechanism that reduces inbreeding. This changes in September when cows become sexually receptive, an interest that lasts in each female only a matter of hours. Optimizing this fleeting opportunity greatly taxes male energy, which is expended in rounding up harems and driving off rivals. Typically, a bull may claim fifteen to twenty cows. Sexual urgency occasionally even leads a bull to go after domestic cows, a dis-

Elk frequent riverbanks; their tracks, droppings, and signs of foraging are evident even when the animals themselves cannot be seen. Antlers weigh up to 40 pounds per pair, a remarkable load that bulls gain and shed annually. Ritualized fighting among bulls—seldom more than a sparring and shoving match—helps establish breeding dominance.

ruptive effort that creates more puzzlement than reciprocity. Minnie Peterson, born on a forest homestead, remembered:

> If cows don't go where he wants, why the doggone bull will take and gore them. You can't do much about it, either. An elk bull, he'll come after you—and come to kill. Years ago people had quite a time. They'd be treed by elk. Bulls kept them up there I don't know how many hours. And you don't want to climb down. That bull will come right back after you.

To study the effects of elk on vegetation, researchers decided in 1980 to lay out two study plots in the South Hoh Valley, expanding on work begun fifty years earlier in the Hoh, Queets, and Quinault. The early work had centered on small plots that were fenced to exclude elk and permit observation of what happens to the forest understory when it is not browsed. The new work followed the same approach but involved surveying two plots of 2.5 acres each—twenty times the size of their predecessors—one situated on a grassy flat near the river, the other on an upper terrace with typical spruce/hemlock forest. The researchers meticulously counted and measured all plants within plot boundaries and identified them by species; then they fenced half of each plot to exclude elk and left the other half open.

Results are essentially the same in all plots and all valleys. Without browsing, the forest's naturally open understory quickly becomes a thicket. Inside the exclosures salmonberry burgeons into six-foot

Opposite
Herds crossing a river pick shallow places with secure footing, the same qualities chosen by human hikers.

Fenced study plots, called exclosures, dramatically demonstrate the effect of elk on forest vegetation. Grasses dominate outside this fence in the South Hoh. Inside—safe from elk—shrubs, ferns, and young hemlock quickly take over. Mosses also increase inside an exclosure, perhaps because they are not trampled by elk. The photo was taken nine years after the study plot was fenced; originally the area inside the fence was as grassy as that now outside, which is still browsed.

Bracken ferns unfurl their coarse fronds as spring growth pushes up from fleshy rhizomes. Native Americans used the fronds to lay fish on and for wiping off scales. They dug and roasted the rhizomes; starchy foods were less abundant for Northwest Coast people than proteins and fats.

Skunk cabbage brightens peninsula lowlands in early spring, but it is common in the rain forest valleys only in the Quinault, infrequent elsewhere. Elk, deer, and bear eat the plants despite sharp calcium oxalate crystals in their leaves. Beetles feeding in the flowerlike structure bring about pollination.

bushes, although elk find it so irresistible that outside the fences it exists only where beyond their reach. Similarly, red elderberry, thimbleberry, and huckleberry become prolific inside the fences, but are robust outside them only where inaccessible to elk. Huckleberry bushes inside the exclosures develop stems conspicuously thicker than those outside, and saplings—particularly hemlock—shoot up.

The lower South Hoh plot displays particularly dramatic differences inside and outside its exclosure. Deer ferns inside the fence grow twenty times as high as those outside. Lady ferns stand waist high inside, although they are so tiny outside that they barely show within the turf. Oxalis, beadruby, and liverworts are twice as large inside as outside. Skunk cabbage with normal three-foot leaves grow inside, mere inches from miniature, heavily browsed skunk cabbage outside the fence. Grass is almost nonexistent inside, but is the dominant ground cover outside.

Clearly, without grazing, what have been meadowlike carpets of mostly grasses become dominated instead by ferns and shrubs. The sheer mass of living plant material increases, but its variety decreases because certain species compete so successfully when protected that others disappear. The exclosures demonstrate that elk do far more than simply live in the forest. They shape its openness, and that openness in turn provides them with succulent forage, which permits faster digestion and therefore more ingestion. Without the grassy meadows the elk themselves create, the animals would have to rely on the coarse browse of the forest, which takes longer to digest than succulent browse and so makes the gut feel full. The animal is therefore less often hungry, eats less, grows less, has less stamina, breeds less, and produces smaller calves less likely to live. Instead of this happening, the act of browsing and the availability of the browse have attained dynamic balance. Food is neither too available or nutritious nor too limited; it is "just right."

This attunement supports an elk population estimated at almost three hundred in the Hoh and South Hoh valleys inside the park and over two hundred in the Queets Valley. For the entire west side of the peninsula the total figure is perhaps 3,500. The elk have been present for at least three thousand years, as indicated by bones of that age found in archaeological deposits at Hoko River, near Neah Bay. The animals' barks and squeals while calling to each other express the essence of this environment: rain forest understory and elk are two parts of a single whole.

The Forest Floor

SOME FOREST-FLOOR DWELLERS ARE OBVIOUS. Winter wrens rustle in the duff like small brown mice, then perch on a twig or stump and pour out a clear, wavering trill that seems to come from Pavarotti lungs squeezed into a three-quarter-inch chest. Raccoons, otters, and skunks are heavy enough to leave footprints in mud and sand, so their presence along streams and riverbanks is clear even if the animals themselves are not seen. There may also be tracks of coyotes, animals now infiltrating the niche that once belonged to wolves,

Townsend's chipmunks, closely related to squirrels, burrow in the ground for shelter and feed on seeds, berries, insects, pupae, and worms. They immediately eat perishable food but carry other foods to cache in rotting logs or their burrows.

OPPOSITE TOP
In undisturbed forest, mammals occupy various niches without much competition. Black bears (top) are omnivores relying particularly on plant food but also eating small animals, carrion, fish, and insects. Cubs stay with their mothers until they pass age two. Raccoons (bottom left) are nocturnal, hunting especially along quiet streams and pools for frogs, fish, salamanders, and aquatic insects. They also catch mice and raid the nests of forest-floor birds.

Shrews (bottom right) are miniature but mighty predators. Three inches long, with a high metabolic rate, they must daily devour food in excess of their own weight.

and of bobcats, one of the smallest wildcats in North America (about fifteen pounds). Elk and deer prints are certain; black bear and cougar, possible.

Most small mammals, however, leave little evidence despite being both numerous and active. A large population of shrews inhabits the forest floor, but they are rarely noticed. Scarcely two ounces in weight, they scuttle about the forest duff, pointed snouts constantly aquiver, hearts thumping at 800 beats per minute. Their metabolism is set so high it can be sustained only by devouring beetles, larvae, spiders, and carrion virtually nonstop and augmenting them in almost any way possible. Water shrews, for instance, swim like miniature otters and capture frogs bigger than they are by biting into the victim's leg and hanging on until they can make a lethal chomp into its skull. Other shrews prey on the forest's notorious, slithering slugs, then may spend an hour struggling to clear mucus from the mouth—potentially a serious problem for a creature needing to eat so constantly.

The outlandishly large slugs are a point of joking pride: they appear as Northwest tee-shirt designs, on postcards, and as the focal point of summer festivals featuring slug weigh-ins and slug races. Banana slugs, perhaps the largest in the world, average six inches in length, although a champion fully stretched out may be twelve inches long. Mucus protects their skin from bacteria and mold and also provides a highway that allows "speed" records: the fastest was officially clocked at 0.034 miles per hour, although one-tenth that rate is more

TOP
Eye spots at the end of banana slugs' long antennae distinguish light from dark and also detect the infrared spectrum; short antennae provide a sense of smell.

MIDDLE RIGHT
At one state of their cycle, slime molds contract, preparatory to producing spores.

ABOVE AND LOWER RIGHT
Sword fern—like all other ferns—propagates from spores borne on the underside of fronds and distributed by air currents.

Liverworts, relatives of mosses, form mats on decaying logs and mineral soil and also grow epiphytically in trees. Snails (upper right) are less noticeable in the rain forest than their huge kin the banana slugs, which are classified as snails that have lost their shells through the course of evolution.

routine. Mucus also lets one slug recognize the trail of another and decipher the direction of travel and whether sexual pursuit is worthwhile. Slugs are hermaphrodites; each has both male and female sex organs and is capable of self-fertilization if no mate is found. Even so, a potential pair may follow one another and cavort for a day; then copulate nonstop for another two or three days and nights, exchanging sperm reciprocally. The penis—an inch long—is so large and turgid that withdrawal apparently is difficult. The solution sometimes is to gnaw it off. Nobody yet knows whether regrowth occurs.

More peculiar, from a human perspective, are slime molds, lifeforms with the oozing locomotion of slugs and the spores characteristics of reproduction in fungi. In its crawling phase, a slime mold looks like a conspicuous blob, sometimes veined and ruffled along its leading edge, sometimes with a knobby texture like cauliflower. Many species are yellow, but some are pink or pale purple. Sizes range from microscopic to three feet across. Slime molds sometimes emerge into the open, but they live primarily beneath leaf litter or inside rotting logs where they feed by engulfing fungi, bacteria, and various microorganisms—and are themselves eaten by slugs. Eventually they send up stalks that produce spores, which are distributed by wind or rain or feeding insects.

If the spores of some slime molds land in a damp environment they develop into amoebalike cells, which move by constantly changing shape, and if they land in water they add threadlike flagella and

LEFT
Extensive beds of vanilla leaf carpet parts of the forest; creeping underground stems allow the plant to spread.

RIGHT
A tree, newly recruited to the forest floor, bridges from one forest generation to the next, providing a reservoir of nutrients and moisture to nourish new life.

propel themselves by whipping their way across even a film of water. If conditions reverse, cells in water lose their flagella and become amoebalike; similarly, amoebalike cells in a merely damp environment add flagella and take up swimming if their minuscule world turns watery. In time the two types of cells fuse and form the mobile slime that gives the organism its name.

Many organisms such as slime molds fit imperfectly into either the Plant or Animal kingdom, a situation that recently led taxonomists to add three additional kingdoms. This new system places bacteria, cyanobacteria, and slime mold in the kingdom Monera; protozoa and algae in Protista; and fungi, yeasts, and molds in the kindgom Fungi.

The plant life of the forest floor is familiar and almost commonplace in comparison with such life-forms as slugs and slime molds, which humans perceive as odd. Ferns stand waist high. Feather moss forms a half-plant, half-air carpet three inches thick. Vanilla leaf floats triple "butterfly-wing" leaves from foot-high stems, holding each leaflet flat to optimize its exposure to light. Nurse logs lie everywhere and have the greatest visual impact, and greatest ecological story, of the forest.

The logs are strewn like gigantic pickup sticks, cumulatively weighing nearly fifty tons per acre. Not waste, as once supposed, this massive presence fills multiple key roles. Lyle Cowles, a National Park Service trail crewman who wrote a weekly series for the *Forks Forum* during the 1960s, once noted: "In trail work I find it necessary to dig

In moist coastal forests, snags that still retain a sound base usually transform in less than a century from a stage of advanced decay into a low stump. This contrasts with a fallen tree, which typically takes about 150 years for its nutrients to be recycled by decay organisms and another 250 years for it to be incorporated into the forest floor. In a Decay Class 3 log, exposed wood is softened although large sections remain firm; bark is sloughing off; and seedlings have joined the mosses that pioneer a Class 2 log. By Decay Class 4, bark is gone; blocky sections of soft wood remain; and vegetation includes shrubs and conifer saplings as well as seedlings. By Class 5, the log is completely soft and nurseling shrub and tree roots have penetrated to heartwood.

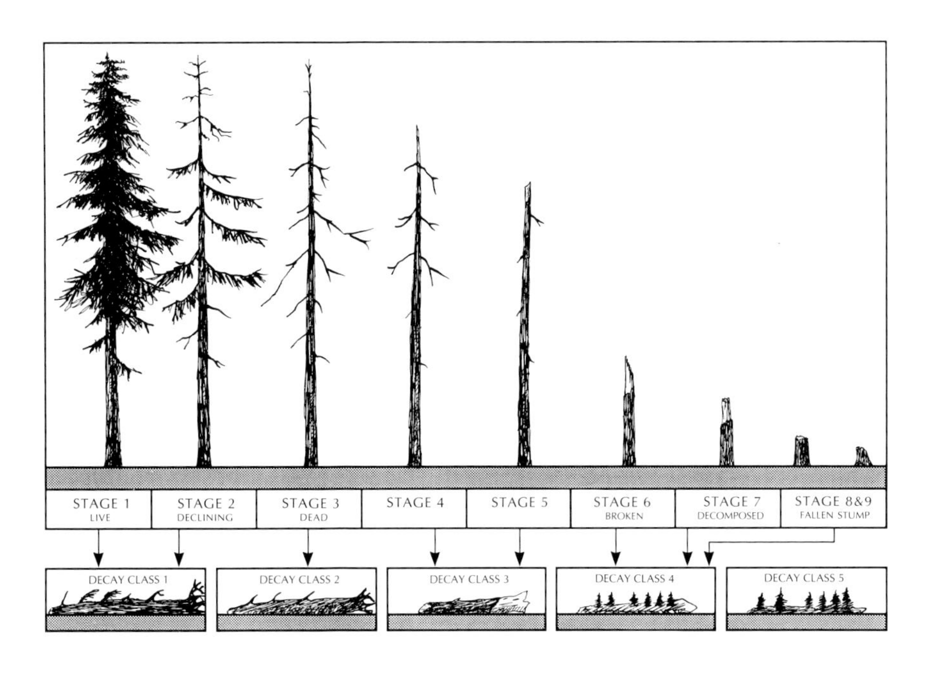

Researchers plotted live standing trees, snags, stumps, and logs during a South Hoh study. They labeled each log with abbreviations of its scientific name and decay class and numbered standing trees consecutively. Each small square represented here measured one meter (39 inches) on a side. Aside from the scientific detail, the drawing illustrates the extraordinary amount of living and dead wood within the forest ecosystem.

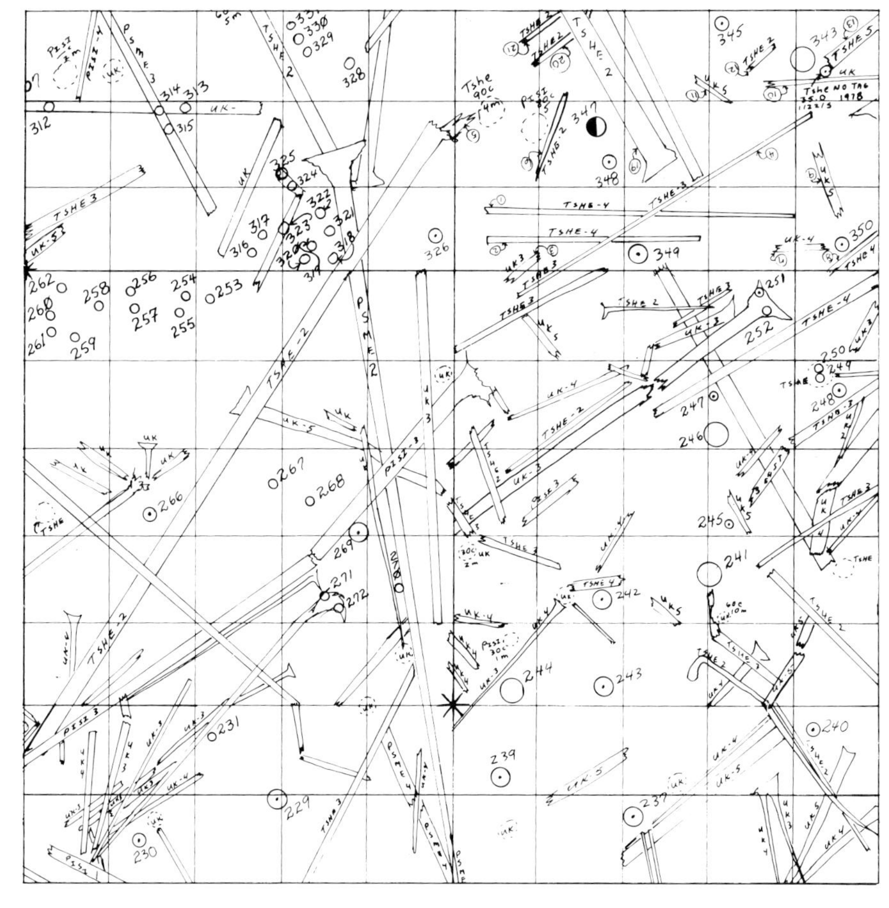

HOH RIVER TERRACES

PSEUDOTSUGA MENZIES II	15-25 cm	LOG
TSUGA HETEROPHYLLA	25-50 cm	STUMP
PICEA SITCHENSIS	50-100 cm	POST
THUJA PLICATA	100-200 cm	
ACER CIRCINATUM	>200 cm	

Garter snakes are the forest's only common snake. This particular species is largely aquatic and usually swims away when startled rather than hiding. It feeds on fish, tadpoles, frogs, and salamanders.

through these decayed trees quite often—and the things I find!" He then listed:

- salamanders ("when dug out they lie quite still if it is chilly, or scurry like mad if it is warm")
- batches of eggs ("about pea size, covered with a translucent membrane and full of a cloudy liquid with two black spots [eyes] floating in it. . . . They must be salamander eggs but seem too large for a salamander to lay.")
- yellow jackets [wasps]
- a nest made of grass ("about baseball size . . . some of the grass still green, probably the nest of a white-footed mouse")
- spruce cones ("a large cache that must be a squirrel's pantry")

Logs lying crosswise on a slope accumulate soil and litter on their uphill side, an effective erosion check. Their downhill side harbors small creatures from shrews and mice to garter snakes and alligator lizards (the only lizard of coastal Northwest forests and decidedly lethargic and plain compared with its desert cousins). Logs also act as perches for grouse drumming their deep, accelerating series of booms that sound like a generator starting up and finally catching hold (a sound produced by air rushing through their feathers, not by wings striking the log). Logs offer ready-made runways above the tangle of the forest floor, as the frequent presence of animal droppings indicates, or sometimes logs thwart travel. After a 1921 blowdown, Hoh

A ruffed grouse perches on a log to sound a courtship message by beating its wings. Logs also provide runways, barriers, and shelter for any wildlife that can crawl beneath them or into pockets of interior decay.

Ensatina salamanders prefer soggy, large-diameter, well-decayed logs as places to lay their eggs, which then swell with moisture.

A snag that is already well-decayed when it falls will become part of the forest-floor structure, but it is already too rotted to serve as a nurse log.

Valley settlers reported massive tangles that trapped bands of elk, whereas deer, which are smaller, were little affected; they could successfully work their way through the maze.

Death is not an end but a change of role for a rain forest tree, a pivotal point on a long continuum. Douglas fir grows for three or four centuries and decomposes for an almost equally long period. Spruce and hemlock decompose in a little more than one century. Perhaps only 10 percent or less of a mature tree's mass is alive at any given moment in any case. The rest is dead wood, which provides structural support. In terms of total cells, needles amount to about 3 percent of a living tree, inner bark and the layer just below it to 5 percent, and ray cells in sapwood to 2 percent. Ironically, a tree may have a much greater ratio of living matter to mass after it dies than while alive. A dead tree actually gains life, from yellow jackets and mice and salamanders to hundreds of feet of plant roots and fungal strands, which penetrate as decomposition progresses.

Inner bark and cambium are the first part of a log to disappear, consumed within about a decade by insects and microbes. Bark beetles and wood-borers that depend on these perishable tissues for food supply and habitat have only a fleeting window of opportunity; they like them while fresh, which means for no more than the first year or so. Microbes complete these tissues' decay.

Sapwood, rich in carbohydrates, also attracts wood-borers. Some enter after feeding on the inner bark and cambium; others go

Opposite top
These relatively young trees blew down only five to ten years ago; their bark remains intact except where scuffed—probably by hikers—and other than moss, the only vegetation is on their upturned roots, which still hold some mineral soil. There, salmonberry thrives, safe from elk. By falling, such trees open the forest floor to light and stimulate the growth of understory trees. The logs will fill as many ecological functions as the trees did while alive.

Opposite lower left
Buprestid beetles are among the wood-borers that tunnel into sound wood, thereby creating surfaces that soon are colonized by decay microbes and fungi.

Opposite lower right
Woody conks of bracket fungus last for years, annually producing new spore-forming tissue on the underside.

A seed coat still clings to a newly germinated spruce that so far is competing successfully with moss on a nurse log.

directly. Ambrosia beetles, among those that chew directly to sapwood, are unusual in that they bring along fungal spores carried in special abdominal cavities. The fungus grows and supplies the beetles with food, although an optimal relation between the two depends on exact conditions. If it is too dry, the fungus fails to grow and the beetle then starves. If it is too moist, the fungus grows too well and smothers the beetle.

Bacteria also arrive with beetles, at first primarily those that metabolize wood sugars, later those that can break down more resistant woody compounds like cellulose and lignin. Some of the bacteria convert atmospheric nitrogen into substances useful to other organisms. Other bacteria contribute to the eventual formation of humus. Additional fungi also arrive, some of them producers of antibiotics that suppress bacteria, others producers of ethylenes that stimulate bacteria. New invertebrates enter the log as well, among them specialists that feed exclusively on bacteria, insects, droppings, or wood. After death, their bodies incubate additional fungal spores and bacteria. Entire communities develop. Nitrogen and phosphorus within the log increase about threefold owing to this activity. The whole process is decay, a coming full cycle. Green plants produce complex substances from simple ones. Decay organisms produce simple substances from complex ones.

To study how logs decompose, Mark Harmon, professor of forest sciences at Oregon State University, has begun a project that is intended to take two centuries to complete. It is part of the Long Term Ecological Research program sponsored by the National Science Foundation—and his project is considered long even within that framework. The U.S. Forest Service's H. J. Andrews Experimental Forest, located southeast of Corvallis, Oregon, is the setting for the study. There, in 1985 Harmon directed placement of some 500 logs within a half-dozen undisturbed forest stands. Each log measures about two feet in diameter and twenty feet long, and is either Douglas fir, western hemlock, red cedar, or Pacific silver fir.

Cut as part of a timber sale and brought to the study sites, the logs were placed at angles to the slope and with various exposures to light, as would be representative of natural conditions. Results were immediately surprising. For instance, rain quickly leaches out far more nutrients than had been expected. Also, insects arrive remarkably quickly. About two days after positioning the logs, Harmon covered several with netting to permit comparing their decay rate with those of logs unprotected from colonization by insects. A week later, however, he found that the *ex*closure nets were actually *en*closures. Ambrosia beetles had entered the logs during the two days before the protective nets were added. Decay begins that quickly.

As the chain of grazing, preying, parasitizing, and scavenging develops within a log, an additional chain starts on its surface: plants cover the top and sides with a green shag rug of oncoming life. The first plants sprout in organic litter caught by bark crevices, usually only to die when the litter dries. Low mosses also begin to grow, un-

Spore capsules rise on slender stalks above the common moss *Leucolepis menziesii*. Space on logs is as much in demand as on the forest floor.

troubled by the log's inhospitality because their nourishment comes entirely from precipitation, dust, and splashing raindrops, which on a minuscule scale redistribute surface nutrients. The mosses need the log only as a perch, but they considerably improve the survival odds of seeds delivered by air currents or brought by birds and squirrels; lodging among these mosses gives seeds more likelihood of finding moisture and a secure toehold. Later, taller mosses are a problem.

A "root race" now begins. Seedling stems lengthen slowly, but root growth is rapid. A sub-bonsai treelet an inch high with a mere rosette of needles may have an eight- to ten-inch root, an umbilical cord linking today with tomorrow. Some roots will intersect the entry tunnels of insects and follow them into the moist and nutrient-rich interior of the log. Rotting wood in this forest doubles or triples its moisture content in fifteen to thirty years, a wet legacy often crucial for seedlings. The wood acts like a sponge left out in the rain, and its decay organisms also produce water as a by-product of their digestive chemistry.

Once decay has prepared an adequate seedbed, the tiny trees tufting a log may stand ten, twenty, or even fifty to the square foot. If the log is within reach of elk, almost all surviving seedlings will be spruce, not hemlock, which the ungulates find irresistible. On higher logs and stumps hemlock grow well. Herbs and shrubs also share the nursery according to its accessibility for elk.

A dozen years after starting, only about 10 percent of the original

Opposite top
Spruce seedlings line a nurse log that is beginning to lose its bark; the next stage of decay offers better growing conditions.

Opposite bottom
An elevated log hosts a first crop of moss and seedlings, which are mostly hemlock. Rooted in crevices, many will fall off as the bark sloughs; the sloughing will fatally expose the roots of others to air and sun. They will quickly be replaced by new mosses and seedlings.

Above
Successful spruce saplings, which survived severe competition while in the seedling stage, flourish on a nurse log. Elk browse hemlock more heavily than spruce, but their role is not a sufficient explanation for differences in the regeneration of the two species. If it were, the low elk population from 1890 to 1910 should coincide with a disproportionately high hemlock regeneration, but researchers have found no such evidence during that period.

seedlings will remain. A hundred years later, perhaps one out of each ten thousand will still be alive. Several factors dim survival odds. Many seedlings fall to the competition of cohorts. Others are lost to the fluffy feather mosses that replace the pioneering mosses. Feather moss holds germinating seeds too high to permit root contact with the log and it cuts the already meager light available for photosynthesis. Sloughing bark exacts another major toll of nurselings. For a Douglas fir this happens after about thirty years; for hemlock, twenty years; for spruce, ten. Only saplings that have reached heights of four to five feet are likely to be rooted well enough to resist physical dumping as bark sloughs or, surviving that, to endure loss of the bark's insulation for their roots. Wet snow may overweight saplings and peel them off the log. Successful neighbors are likely to crowd some off.

Such eliminations are inevitable. Space is finite: scores of beginners fit well where later there can be only one. A tree that survives these early vicissitudes has only a few more years to get its roots to mineral soil. Plants germinated on high stumps or logs not resting on the ground may survive for the first thirty to forty years while rotting wood offers enough nourishment, particularly for saplings that make successful root contact with certain kinds of fungi. After that, the stability of soil is needed. Therefore the first nurselings to succeed in sending roots into the ground win the race for survival and become the potential next generation of forest monarchs. Even in old age they are characteristically still lined up with one another, testimony to the

Standing Douglas firs occasionally host nurseling trees. Their rough bark, which is firmly attached, catches and holds litter.

Roots seem to prop trees like this hemlock, but their form actually reflects beginnings on a nurse log or stump, not a genetic trait of buttressing.

Old-growth spruce trees stand in a straight line marking their origin on a nurse log. Such colonnades are a hallmark of the Olympic rain forest.

nurse log that gave them their start. Such trees form colonnades, a prime signature of this forest. Their tentaclelike roots continue to grip the remnant hollow of their log even after it has moldered and disappeared. Mentally measure the diameter of this hollow and translate the figure into the probable age of the tree when it fell; then add the size of the largest tree in the colonnade, also translated into years. The sum of these two ages bridges from the present back to when that nurse log began life on its nurse log—six, eight, ten centuries ago. The process has kept the forest cycling and recycling for millennia.

Below Ground

HUGE TREE SIZE IN THIS FOREST is directly linked to the unseen realm within the soil, where water droplets like microscopic seas swarm with bacteria and fungal strands (mycelia) form jungles of pale growth. Without minute, below-ground life-forms and interactions there would be no forest as we see it today.

Perhaps the most effective of these interactions are mycorrhizae, literally "fungus-roots," a type of union apparent on the earliest-known fossil roots, hundreds of millions of years old. Some fungi directly enter root tissues, others mantle the surface. Both types of colonization produce an essentially new structure, a fusion of tree roots and fungi into a new form: mycorrhiza. A century of study clearly indicates the crucial importance of the relationship, although it only partly clarifies the complexity. In greenhouse experiments decades ago researchers injected certain fungal cultures into Douglas fir seedlings to test the effect of fungi on roots. No other factor was changed, yet

Macrotyphula (top left), a decomposer fungus, grows on cedar litter; *Clavaria* (lower left) probably is mycorrhizal although which species it colonizes is not yet known. *Geastrum* (right) is a decomposer and possibly also mycorrhizal.

the seedlings spurted rapidly from bare survival into normal growth. Later work has shown that roots, even of the same tree, look entirely different if they are integrated with a fungus. Mycorrhizal roots are more swollen and branched than those without fungi.

All mycorrhizae absorb water, minerals, and nitrogen from the soil and transfer them to trees. Many also stimulate the growth and life span of root tips by releasing enzymes that free various nutrients otherwise chemically bound and unavailable. In addition mycorrhizae ward off tree diseases by excreting antibiotics, stimulating other organisms that inhibit pathogens, and forming barriers that physically thwart the entry of pathogens. Furthermore, they secrete substances that bind soil particles and enhance pore stability, which is beneficial to the subsurface movement of water and air. In return for such services, fungi derive energy from carbohydrates transferred from the tree's foliage to the roots. They could not live independently. Thus, giant spruce or hemlock reveal more than today's trees and yesterday's nurse log: they are the product of an unseen linkage between radically different organisms, each dependent on the other.

An immense amount of fungal material is present in the forest floor. Soil of old-growth Douglas fir forests has been found to contain nearly two tons of fungal mycelia (dry weight) and two and a half tons of mycorrhizae per acre. Some mycorrhizal fungi readily associate with a range of hosts; others are limited to one particular plant species. Some fungi living on root surfaces actually link separate plants below ground. Their mycelial networks are capable of transferring

Fungi vary greatly in form and ecological role. Some secrete enzymes that break down dead tissue. Others form mycorrhizal links with roots, greatly increasing their hosts' ability to absorb moisture and nutrients. The relationship can be crucial. For example, hemlock seedlings live only about a year unless their roots are colonized by fungi. The drawing shows fungal hyphae of above- and below-ground fungi contacting roots; the resulting mycorrhizae (right circle); a non-mycorrhizal root with roothairs (bottom circle); and a red-backed vole eating a truffle (left circle).

Flying squirrels glide from tree to tree and to the forest floor where they nibble above-ground mushrooms, berries, and carrion and dig below-ground truffles. Their skin flaps double the undersurface of their bodies, helping to buoy them. The squirrels seem to stay almost constantly in motion through the night, ranging as much as a linear mile and rustling so noisily that spotted owls can prey on them simply by perching and listening in a known feeding area.

carbon from one plant to another, even—astonishingly—from one species to another.

Fungi familiar to most humans—mushrooms and puffballs—are spore-producing organs pushed above ground, largely dependent on air currents to waft spores to sites suitable for growth. Other fungi—the truffles—belong entirely to the subsurface realm. Their spore-producing organs, which look like small ill-shaped lumps, are valued worldwide by people from nomadic Kalahari Bushmen and Australian aborigines to European and Japanese gastronomes. Nor are humans alone in valuing truffles. In the Olympic rain forest and throughout Northwest coastal forests, rodents dig truffles. They are drawn by odors, which vary by species from cheesy, spicy, or fruity to fishy and outrightly foul.

The extent of the relation between mammals and below-ground fungi became apparent in the 1980s after mycologist James Trappe, of the U.S. Forest Service's Pacific Northwest Research Station, joined zoologist Chris Maser, then with the federal Bureau of Land Management, for a one-day field investigation at Wheeler Creek Research Natural Area in southwest Oregon. Maser had with him several vials holding stomach contents from rodents collected as part of other research. Trappe previously had noticed deer and elk pawing out truffles, and he had regularly used the small pits and tailings left by truffle-seeking rodents as a guide to finding them himself. The idea of checking the vials of stomach contents for spores occurred to both men. Were major mycorrhizal species of fungi eaten by rodents? By which species? Do spores remain intact during digestion?

Trappe began the identification process, an undertaking soon known to friends and colleagues who sent him "reeking boxes with plastic bags of unspeakable material." Hundreds of examinations later, he could say that flying squirrels, red-backed voles, and deer mice feed on mycorrhizal fungi, as do chickaree squirrels, chipmunks, and hares (and also creatures from armadillos to wallabies, according to subsequent worldwide research). Indeed, truffles are a major part of the diet among many animal species whose fecal pellets are a prime means of inoculating tree roots with fungi. More than half a million spores may be in a single pellet, five hundred times the number needed to colonize a root tip. Ironically, commercial forest managers try to eliminate rodents from new plantations because their nibbling destroys some of the struggling young trees, but their feces actually contribute to successful growth.

Recent research has established that below-ground relationships also involve certain bacteria and yeasts that pass unscathed through the digestive tracts of rodents. Among the bacteria are some that feed on a substance exuded by fungi and reciprocate by converting soil nitrogen into a form usable by both the fungus and the host tree. The yeasts fit into the linkage by producing compounds that nourish the bacteria and stimulate fungal growth. The whole mix seems to benefit from being chemically processed in a special intestinal pouch of rodents, where fungal spores, yeast, and bacteria are concentrated prior

RIGHT
A flying squirrel peels back the tough outer sheath of a truffle to reach its flavorful inner flesh.

BOTTOM LEFT
Truffles are the below-ground spore-producing structures of fungi like *Hysterangium coriaceum*, shown here with its threadlike hyphae colonizing a tree rootlet.

BOTTOM RIGHT
Hysterangium spores pass unharmed through rodents' digestive tracts and are thereby widely distributed.

OPPOSITE TOP
Plant roots and fungal mycelia form a mutually beneficial partnership, illustrated by this microscopic view of a mycorrhizal root cross-section. Hormones released by the fungus stimulate root branching, which increases absorptive surfaces.

OPPOSITE BOTTOM
Mycorrhizal fungi mantle small rootlets and grow between cells to form an interlocking network. Sugars produced by the plant and transferred to the roots nourish the fungus.

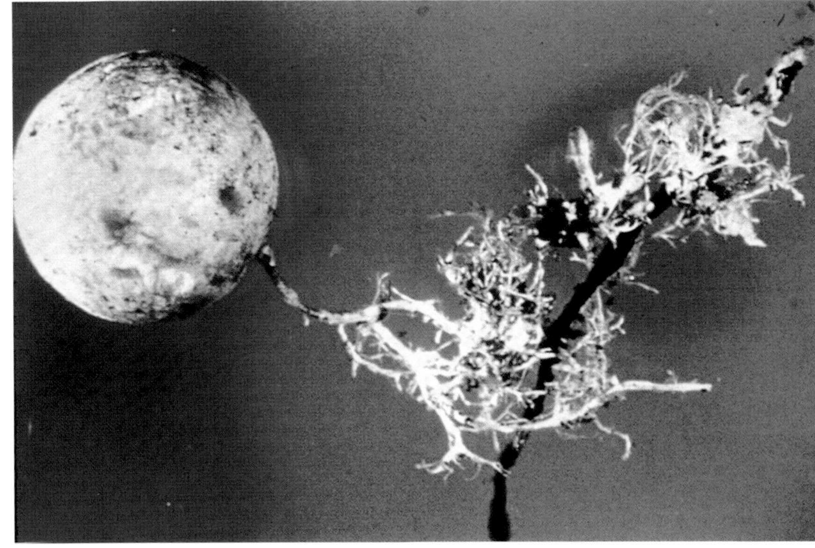

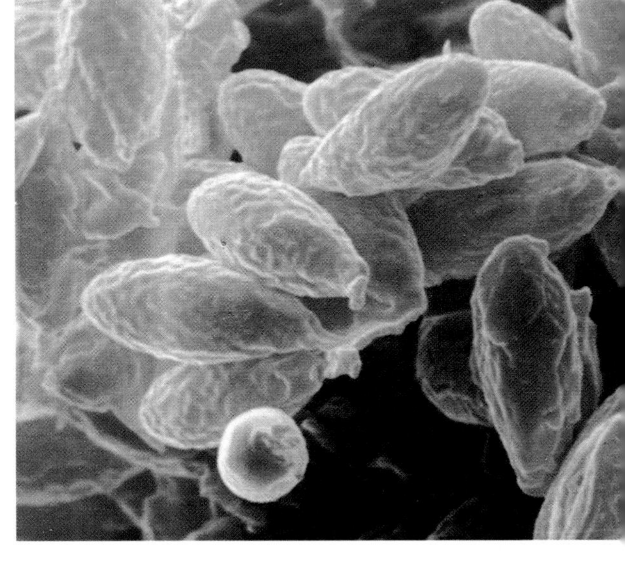

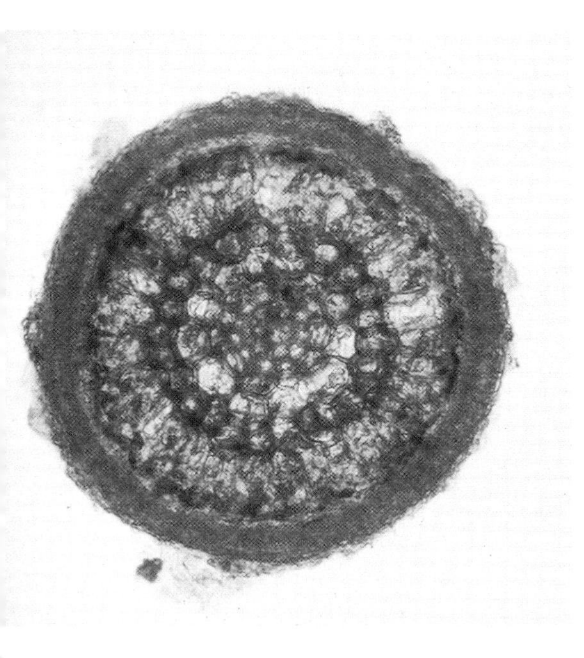

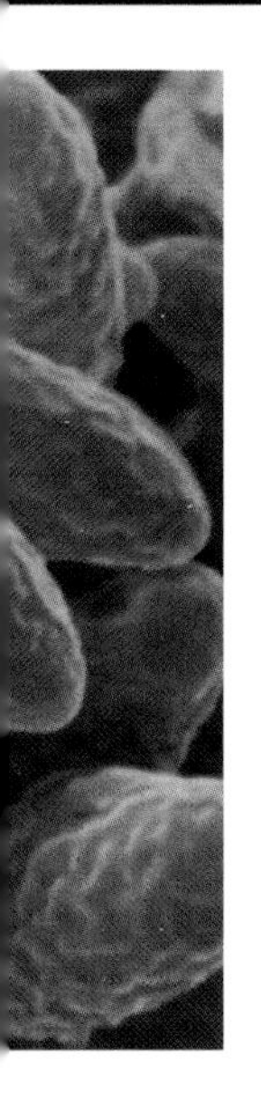

to excretion in feces. The chain is from yeast to bacteria and fungus, to trees, to mice, squirrels, and voles, and back again to fungus. Furthermore, this chain functions within a realm also populated by uncountable quadrillions of free-living bacteria and hundreds of thousands of mites and nematodes for every square meter of surface, plus worms, centipedes, insects, and burrowing mammals. The mammals often sever roots while digging their tunnels, causing new root tips to form. New roots are ideal sites for the growth of fungi, which are present as spores in the feces of the mammal that did the burrowing and cut the root.

Such interactions are dynamic. Only about one-fifth of forest biomass is located below ground, yet that portion may use half or more of the total energy produced by photosynthesis. Trees replace their fine roots two or three times faster than their foliage. Soil nourishes trees, and trees reciprocate by transferring energy in the form of sugars and other carbon compounds from the canopy to the ground. These substances are released by the decomposition of dead roots, by leaching or exuding from live roots, or by the grazing of minute invertebrates that feed on root tissue. In one study roots were found to return potassium to the soil at a rate triple that of its above-ground return in leaf litter and twigs.

Such relationships have evolved through time. Trees need different kinds of mycorrhizal fungi at different stages of life and at different levels beneath the ground. Furthermore, fungi that fruit below ground apparently are descended from those that fruit above ground; below-ground fruiting is more efficient and offers a favorable trade-off. In exchange for developing odor signals to attract animals and acting as a nutritious, moist food source, below-ground fruiting protects fungi from heat, drought, and frost and frees them from depending on random air currents for the dispersal of spores. Instead their spores are optimally packaged for effective dispersal. At least one Northwest mammal, the California red-backed vole (not found north of the Columbia River), seems to have evolved as an underground-fungus specialist; its frail skull and tooth structure are suited only to soft foods, and its menu is restricted almost exclusively to truffles. Flying squirrels also rely on truffles at certain seasons—and while digging out their dinner are likely to themselves become food for predators from spotted owls to bobcats and coyotes.

Such a web of interactions points to the likelihood that today's forest developed not as individual species living together by chance, but as an ecosystem, a whole. The unseen realm beneath our human feet belongs to the unity of which we, too, are a part.

Rivers and Terraces

Rivers in each of the rain forest valleys are the main arteries of the forest community, not mere liquid conduits. They are dynamic biological systems and the primary physical force that sculpts the bottomlands and produces various surfaces from forested terraces to gravel bars, formed as today's active channels braid back and forth across their wide beds.

Gravel bars form as channels braid back and forth across wide riverbeds, responding to changes in water flow. The bare, gravel-floored beds of the Hoh, Queets, and Quinault rivers occupy only about 10 percent of their valley floors. Sequential terraces carved by the rivers during thousands of years provide sites for the development of both forest and wetland ecosystems.

Walking from a valley's sidewall to the river gives a sense of geological and community sequences. The walls are steep and relatively dry; therefore Douglas fir and Pacific silver fir, rather than Sitka spruce, codominate along with western hemlock. At the base of the sidewalls are the oldest valley bottom terraces. Their surfaces host spruce and hemlock, the final, climax stage here, which is self-perpetuating unless the environment changes.

The next lower terrace is typically forested with spruce and hemlock intermixed with scattered alders, abundant vine maple, and huckleberry, and carpeted with grass, moss, herbs, and ferns. Native herbs on this terrace and on the alder flats just below are joined by introduced buttercup, self-heal, rumex, redtop grass, and other interlopers that probably arrived when packers started concentrating horse travel on these two terraces in the late 1890s and as ranchers grazed cattle there before establishment of the park.

The alder flats, more recently formed, furnish a transition from conifer forest to river. Large alders dominate but are mixed with occasional northern black cottonwood and young Sitka spruce. At the lower edge thickets of alder and willow saplings clothe surfaces barely above river level, where logs caught on gravel bars slow floodwater and cause sediments to settle out, forming moist, virgin deposits. Winds bring willow seeds, which germinate within hours and produce seedlings able to thrive despite blowing sand, trampling and browsing animals, and recurrent flooding. Roots penetrate to the cobble zone

TOP
Slow, gentle tributary streams provide exceptionally productive habitat.

BOTTOM
Self-heal is a non-native plant found only on lower terraces, one of several species introduced by people within the last century.

beneath the sand, and stems comb floating twigs, leaves, and grasses from water that inundates them, thereby enriching the soil as the flotsam decomposes. Alders join the conquest. Microorganisms called actinomycetes, which have characteristics of both bacteria and fungi, colonize their roots and convert soil nitrogen into compounds usable by plants—the same microbial process as that performed by canopy lichens, leaf scuzz, and fungi in the soil and rotting nurse logs. Much of the alders' nitrogen ends up in their leaves, which contribute nearly two hundred pounds of fertilizer per acre per year to the soil. This enrichment helps to nurture additional plants, some of which replace the alder as forest succession follows its course from pioneering vegetation to climax forest.

Secondary streams typically refill relict channels and swales left long ago as the main river eroded its floodplain, producing today's series of terraces. Streams on the lowest terrace carry primarily overflow from the main channel and water deflected from the head of gravel bars by logs and other debris. Middle-terrace streams carry water from higher terraces and side slopes, and they may also be fed by springs. Their incredible clarity and barely perceptible flow contrast markedly with the rush of main rivers and side channels, which are often laden with silt, and with the foaming white torrents of steep sidewall tributaries fed by melting snow.

The quiet water provides crucial habitat for fish. Food and resting places are available, and tangles of woody debris offer at least

Old-growth forest acts as a sponge, soaking up rainfall and releasing it slowly. Low silt production and effective filtering assure the ecosystem of exceptionally pure water; shade moderates temperatures.

some safety from predators. Coho fry arrive in side streams from their natal gravels in late spring and early summer at about two inches in size, and a second group (nearly double that size) comes in fall and early winter as freshets raise the water level. Some have migrated ten to fifteen miles downriver before swimming into the secondary streams, where they remain until leaving in spring to continue their migration to salt water. The streams, associated ponds, and sedge swamps contribute to the diversity that all fish need at various stages of their lives. Their various conditions attract aquatic species needing habitat away from the tumultuous flow of main river channels.

The zone beneath and along the sides of rivers, hydraulically connected with the flowing water, must also be occupied by aquatic organisms. Researchers elsewhere (specifically in Montana and Colorado) have found that this zone is populated by more life-forms than live in the rivers themselves. Quite surely the same situation prevails in the Olympic rain forest, where rivers braid across wide floodplains with deep alluvial deposits much like those in the Rocky Mountain region.

This zone, where riverbed and forest converge, extends more than thirty feet deep and reaches up to a mile beyond the banks of rivers. Water-filled spaces within the gravel support two types of fauna: specialized invertebrates that spend their entire lives there (chiefly aquatic insects and crustaceans, plus some mites and spiderlike creatures) and others that move temporarily in from the river to take ad-

OPPOSITE
Logs fallen into steep sidewall streams create stairlike steps, which slow and direct the water's flow, catch sediments and organic litter, and produce riffles and pools suitable for life. Swift, plunging streams support less diversity of life-forms than slowly flowing water.

Rain-swollen rivers undercut their banks and topple trees. Such debris is not waste; it contributes to the structure and life of the river.

vantage of stable temperatures and to gain refuge from large predators and protection from floods and droughts. Some members of this fauna are an inch or more in length; most are minute. Several are new to science (study of this zone, called the hyporheic, began only in the late 1970s). Other organisms, including stone fly larvae, are recognized as top consumers of a food chain based on organic detritus. The concentration of nutrients in the hyporheic zone is actually greater than in the river itself; outflow from its gravels apparently enriches adjacent rivers. The hyporheic links life, water chemistry, and geological structure.

Large pieces of woody debris, which are found in channels of all sizes, are a much more obvious but equally remarkable influence on the aquatic nutrient base. Alder breaks comparatively easily, decomposes quickly, and drifts downriver. But large conifers, especially those with branches and roots intact, lie where they fall in small streams, trapping organic litter and holding it in place long enough to be used as food by microbes and invertebrates. Without this woody debris, small streams would lose their nutrient base. In main rivers, fallen trees tend to ride the current, then accumulate along bends and gravel bars. Logjams result and are remarkably stable even in high-velocity rivers unless debris is angled directly into the flow.

Large conifers decompose even more slowly in water than on land because waterlogging prevents oxygen from diffusing through the wood, and without oxygen most decay organisms and invertebrates

A ground-water zone called the hyporheic provides habitat both for specialized organisms found only there and also for some that hatch in water and then emerge onto land. Hydraulically connected with the main channel, this zone may contribute significantly to the nutrient base of rivers. The invertebrates depicted here, left to right, are: bathynella, 40 times actual size; archiannelid, 400 times actual size; blind amphipod, three times actual size.

Benchlike terraces cut by the river characterize the floors of the rain forest valleys.

cannot live. By remaining intact for centuries, the wood offers countless generations of aquatic invertebrates a place to attach eggs, molt or pupate, and emerge into aerial life. Furthermore, it provides surfaces suitable first for algae, then for aquatic fungi and bacteria. These plants soften the wood and create an environment attractive for invertebrates including various beetles and caddis-fly and mayfly larvae. The newcomers cling to the preconditioned wood with bristles, hooks, and suckers, augmented in some species by streamlined shapes that lessen the water's tug. Their rasping roughens the surface of the wood and provides habitat for still more organisms. Gouging, boring, and tunneling insects arrive and open additional surfaces, soon colonized by fresh microbes, which further soften the wood. Larvae of crane flies and stone flies feed on woody debris and leaf and needle litter only after this softening, which may take months. Other invertebrates filter organic matter from the water and harvest bottom sediments. Predators from insects to salamanders and fish complete the chain, feeding on the smaller organisms that are processing the wood.

Wood in water also acts as baffles and dams that alter flow and therefore influence water temperature, how much dissolved oxygen it carries, and how readily invertebrates, amphibians, and fish can avoid getting swept away. The wood produces backwaters and plunge pools, infinite crannies and pockets. It filters debris, catches sediments, and affects streambed particle size by causing water to slow and drop its load. Gravel accumulates in riffles downstream from logs, suitable as

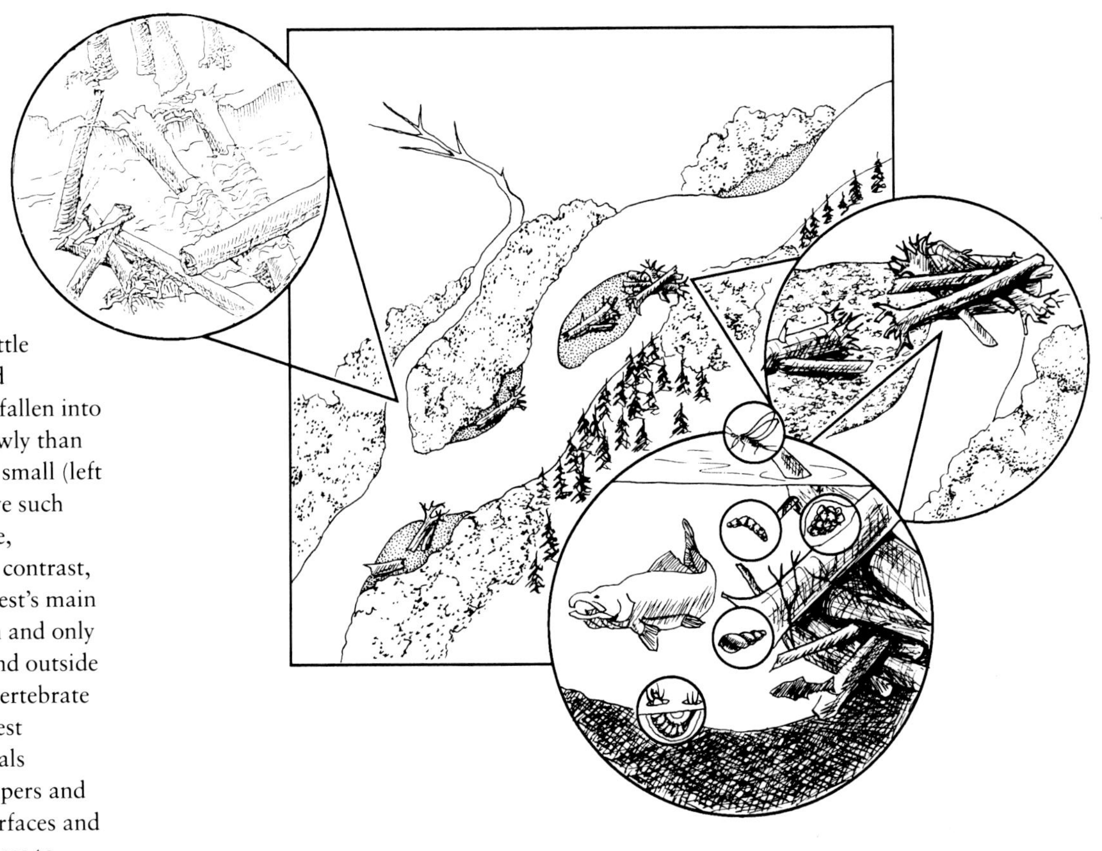

Waterlogged wood has very little oxygen available for fungi and invertebrates. Therefore trees fallen into a stream decompose more slowly than those on land. If the stream is small (left circle), its current cannot move such trees, and they remain in place, providing a stable habitat. By contrast, trees that fall into the rain forest's main rivers are flushed downstream and only those caught on gravel bars and outside bends are likely to persist. Invertebrate shredders and collectors harvest sediments and organic materials concentrated by logjams; scrapers and grazers feed on submerged surfaces and climb onto above-water surfaces to change from larval form into adults.

places for mature salmon and steelhead to lay and fertilize their eggs. For spring/summer chinook alone, forty such spawning grounds can be expected in the two-and-a-half miles of river between the end of the road in the Hoh Valley and Tom Creek. Before excessive fishing pressure by humans, the density of spawning sites may have been ten times greater. Adult coho may have numbered as high as four hundred per mile in some peninsula rivers and streams.

Salmon and steelhead come year-round to spawn: spring/summer chinook from April to August, fall chinook from September to mid-November, coho from mid-September to late January, winter steelhead from November to April, summer steelhead from May to October. Small runs of pinks spawn in the lower Queets and Quinault; chums in the Queets, Bogachiel, and lower Hoh and Quinault; sockeye in the Quinault, where they spend up to ten months in the lake as a transition between ocean life, egg laying, and death. This is an extraordinarily leisurely pace compared with that of coho, which enter the Quillayute River and swim in a single month to spawn in the Bogachiel, Calawah, and Soleduck rivers (the three inland branches of the Quillayute), distances of up to sixty-five miles.

Such diversity of timing and pace is apparently an evolutionary adaptation to the variety within the rivers, a way of filling available niches at sustainable levels. It evens out space for eggs, and lets hatchlings draw from the food chain without overtaxing it. Some salmon stocks are genetically coded for fast growth, others for slow. Indeed,

Opposite top
Salmon differ from all other fish in needing salt water for most of their lives but fresh water to begin life. On the Olympic Peninsula their longest journey from the ocean back to natal water is up the Quillayute River and into the Soleduck, 65 miles.

Opposite bottom
Otters can swim either by paddling with their feet or simply by undulating their bodies. They feed on frogs, salamanders, aquatic beetles, fish, ducks, and carrion.

Using a glass-bottomed viewing tube that eliminates surface reflections, a fishery biologist peers into the water of a logjam where salmon carcasses would be likely to catch. Such carcasses increase nutrients in the river and also are directly consumed by wildlife from otters and shrews to flying squirrels and eagles. The rain forests of the entire Northwest Coast support one of the world's most productive fisheries.

their presence and rhythms are exquisitely linked to the forest as a whole. A female coho lays about three thousand eggs. Three to six hundred of these fish survive to the fry stage; between thirty to ninety of these fry live to maturity and feed in the ocean; and perhaps five or six return to spawn. Most (depending on the particular river system) will be caught before they succeed. The rest will die after spawning. Chinook and steelhead start out with five thousand eggs per female, but end up with the same ratios of survival and spawning as those of coho. These losses at each stage fuel the nutrient base of the rivers through scavenging, predation, and decomposition. The fungus patching the back and fins of a spawner or a stranded fish too weak for further struggle is part of the process by which nature rewrites individual roles. Death is a mechanism for cycling elements from one life-form to another.

Bears carry weakened and dead salmon to the forest's edge to eat and usually leave only lower jaw, gills, gut, and fins when they have finished. Otters hunker on partly submerged logs to feast, and they eat all but skin, backbone, head, and tail. Mink specialize so extensively on salmon carcasses that one early-day Hoh Valley fur trapper placed sets for them according to the times and locations of spawning runs. Such interrelations often impressed explorers and early settlers. Private Harry Fisher, emerging onto the Queets floodplain in the fall of 1890 during an army expedition, wrote in his journal:

> great salmon threshed in the water all night long, in their efforts to ascend the stream. Wild animals . . . snapped the bushes in all directions, traveling up and down in search of fish. At every few yards was to be seen . . . where cougar, coon, otter, or eagle had made a meal.

Nearly a century after Private Fisher's observation, biologists led by Jeff Cederholm, of the state Department of Natural Resources, and Doug Houston, of the Olympic National Park research staff, studied the phenomenon Fisher had noted. Over a three-year period they collected and tagged the carcasses of nearly a thousand freshly spawned coho from hatcheries and placed them a few at a time in small rain forest streams. In their report they described this part of their task as calling for a "hypertrophied sense of humor and an atrophied sense of olfaction." Following placement, the team walked the streams every few days, recording what happened to the carcasses: were they snagged on woody debris? free-floating? buried? caught on a sandbar or in a riffle? lying in an eddy? a pool? Results indicate that most salmon carcasses flush no more than one to two hundred feet downstream, and very few travel more than a quarter mile. (A similar study in the main channel of the Skagit River, north of Everett, indicated that even during a high-velocity flood about half the spawned-out salmon drift no more than six miles, the other half up to twenty-five miles.)

In the Olympic rain forest study, gravel bars served as the final resting place for a quarter of the carcasses; eddies and riffles for lesser proportions. Most became wedged beneath branches and logs. Consumption by wildlife varied from just under one-third of the cumulative mass of all carcasses in certain streams to more than three-quarters in others. Forty-three species of mammals and birds were seen near the streams during the study. Raccoons, otters, shrews, skunks, mice, bears, and bobcats clearly fed on carcasses—as would be expected—and chickaree squirrels, flying squirrels, and perhaps mountain beavers apparently also were attracted to them. Even deer may have at least investigated them because their tracks were found near some of the carcasses, and occasionally elsewhere deer have been observed nibbling dead fish. Eight species of birds fed on the carcasses: ouzel, gray jay, steller's jay, crow, raven, bald eagle, red-tailed hawk, and winter wren.

It is probable, but not yet well studied, that nutrients released from rotting salmon carcasses nourish the microbes that are needed to process litter fallen from the forest canopy; the salmon's substance may diffuse into the food chain and cycle on to aquatic algae and invertebrates and thence back to fish.

Forest and Ocean

SALMON LINK THE FOREST TO THE OCEAN in their remarkable annual journey to spawn. Trees form a less obvious link. Many that topple into rivers wash up along beaches as driftwood. Stranded at the line of high-tide storms, they form a jumbled bulwark of a magnitude unknown anywhere outside the Northwest. Huge logs strewn along the beach stabilize the shore and protect low terraces that

Goose barnacles of the genus *Lepas* depend on floating logs and other flotsam as attachment sites; the barnacles cannot live after washing ashore. A variety of other species attach themselves among the barnacles while the logs are in the water, an example of how forest production influences marine life.

Opposite
More logs accumulate on Northwest beaches than anywhere else on earth. But rivers today wash fewer huge trees from the forest to the ocean, visual evidence of a change that is sure to affect coastal ecosystems.

host beach ryegrass, sedge, beadruby, salal, salmonberry, crab apple, and pioneering spruce. Other logs, still afloat or refloated by high tides, rip loose seaweeds and batter the beds of barnacles and mussel-encrusted rocks. This seeming destruction actually helps to maintain the mosaic of intertidal life, assuring a variety of species and stages. Logs themselves may even become attachment sites, although they are unstable habitats since they often drift off and are repositioned elsewhere.

Logs that escape shore currents float to the open ocean and may eventually enter the great clockwise gyre of northern Pacific currents. Driftwood on some Hawaiian beaches comes mostly from North America (and much of the rest is from the Philippines, Japan, and Malaysia). Hawaiian chiefs particularly prized Douglas fir logs and used them for the huge outrigger canoes that so impressed European and American explorers and missionaries.

Logs distribute wood-borers and sundry attached marine plants and animals while floating and, perhaps because of these food sources, fish tend to congregate around them. This is particularly true of those that float upright with limbs and roots intact. Seine fishermen can increase takes of tuna as much as four times if they set their nets near such logs. Why this should be true is not fully known. It may not be because of direct feeding on attached organisms but because a log's shadow makes zooplankton easier to see. Small fish come to feed on the still smaller fish that feed on the plankton, and they themselves get

Waves break against the small offshore island and long, straight beach at the mouth of the Raft River, just south of the Queets.

eaten by larger fish. Or, instead of a feeding site, perhaps the logs are valuable as objects to rub against to get rid of parasites or as places to attach egg clusters.

The sea bottom itself also is affected by the forest. Trees sunken off the mouths of rivers become focal points for ocean-bottom life. If not driven ashore by waves, these logs usually become quickly infested by gribbles and shipworms, which bore elaborate galleries and excrete wood particles in their fecal pellets. The pellets attract as many as forty species of deep-sea invertebrates dependent on detritus for food. Additional deep-sea species also find the sunken wood and use it for shelter. Certain hermit crabs live in hollow plant stems, and various nematodes and worms frequent the burrows left by borers.

Fine woody debris also enters the ocean, swept miles offshore by rivers and then caught in currents that flow beneath the zone of wave motion (deeper than 130 feet). These bottom currents follow the Olympic coastline north in winter, south in summer. About half the organic matter in bottom sediments, even as much as twenty-five miles offshore from the Columbia River mouth, originates in the forest. No comparable figures are available for the Olympic Peninsula, but from an airplane the outflow from rain forest rivers shows for miles.

With huge trees now rare and fewer and fewer of them in the rivers, will their loss disrupt not just organisms clearly dependent on old growth but also coast cycles?

A glaucous-winged gull uses a drift log at the mouth of the Quillayute River as a perch.

III

TIME AND THE FOREST

We all travel the Milky Way together, trees and men.

—John Muir

FOR THE LAST FIVE THOUSAND YEARS the forest as we know it has been regenerating itself. It existed in this form when Egyptians were building the pyramids near Cairo. Many of the spruce and hemlock trees we walk among were standing when Sir Francis Bacon (1561–1626) and Johannes Kepler (1571–1630) first recognized the value of objective data over mystical portents. These trees have been pushing their roots through the soil and wafting seeds into the air throughout the entire history of science. Individual cedars, which are longer lived, have existed for twice as long as human minds have been free of mediaeval fetters.

What Pollen Can Tell

POLLEN GRAINS ARE SO WELL PRESERVED IN BOGS that samples taken from increasing depths can be read for thousands of years back, inferring vegetation from even before today's rain forest developed. Deposits near Forks indicate that the forest has been coniferous for millennia, although individual species (and genera) have waxed and waned in prominence. Lodgepole pine dominated when the climate warmed 13,000 years ago, closing the Pleis-

Individual red cedar trees are long lived, but the species is a relatively new member of the Olympic rain forest community.

Plants in the rain forest produce prodigious quantities of pollen, which is widely dispersed by wind. It preserves well and can readily be identified at least to genus, often to species. The sudden appearance of red cedar pollen (top left) in Northwest bogs coincides chronologically with archaeological evidence of Native American tools suitable for splitting massive house planks and hollowing dugout canoes.

Lower left
Makah Museum replica of an ancient house buried by landslide at Ozette.

Right
Theodore Hudson, Quileute canoemaker, at the mouth of the Hoh River with his grandson Cliff, about 1963.

tocene Ice Age. Its pollen lies along with that of western and mountain hemlock and lesser amounts of white pine, Sitka spruce, and true fir. Spruce and western hemlock dominate today, and lodgepole pine grows primarily scattered through the coastal forest.

About 8,000 years ago Douglas fir arrived in the Northwest. Just why or where it came from nobody really knows. Its pollen simply appears suddenly within the muck of bogs. Some 3,000 years later, the climate became much as it is today and Douglas fir dwindled, unable to compete in the newly shady environment of what had developed into a spruce/hemlock forest.

About 5,000 years ago western red cedar arrived. Interestingly, its appearance in the pollen record coincides with archaeological indication of a new human cultural pattern. Specialized woodworking tools such as adzes, mauls, and antler and wood wedges are not present in Northwest Coast archaeological deposits from more that 5,000 years ago, but are frequent in more recent deposits. Native American woodworkers—superbly knowledgeable about all valuable resources—evidently recognized the advantages of this new material and quickly developed the technology to use it. They split cedar logs into house planks as long as twenty feet and three to five feet wide, and crafted dugout canoes large enough to hold thirty people with supplies and equipment.

The arrival of cedar completed the roster of principal forest species as we know them today. Changes, of course, continue, following

The ground must have been dry and firm when wind snapped this spruce or it would probably have blown over instead of breaking. Historically, blowdowns of one thousand or more trees occur every decade in Washington and Oregon. On such occasions, trees of all ages, dead and alive, are likely to uproot or break.

a pace that seems gradual though it usually is compounded of abrupt, local disturbances.

The Role of Disturbance

Trees do not all germinate, grow, mature, and simultaneously topple. Natural forest is not a slowly maturing crop. Rather, it is a mosaic of species and forms shaped and maintained through time by disturbance, which constantly restarts the sequences.

In the rain forest, the main results of natural disturbance are trees blown down or washed out by floods, and rocks and sediments deposited at the bases of sidewalls and at the outlets of steep tributary streams. Other forms of disturbance tend to be inconsequential. Outbreaks of bark beetles kill only scattered individual trees. Defoliating insects are rare. Disease organisms are present, but their effects are chronic rather than catastrophic. Drought plays no real role. Occasional summer weeks turn overhead moss drapes oddly dry and change the mud of trails to dust, but the forest does not really suffer. Spruce trees may even open their buds and grow a bit, then set new buds when the drought ends. Heavy winter snow sometimes breaks branches and treetops already weakened by rot, and may cause trees to fall if preceding rains have oversaturated the soil. Wildfires occur at intervals of about six hundred years between one natural burning and the next; they are much more of a factor in Douglas fir/hemlock forest, which is somewhat drier. When fires occur in the rain forest valleys,

Revegetation occurs on the roots of wind-thrown trees more quickly than on rotting logs because their large surfaces remain stable and mineral soil clings to the tangle.

flames tend to stop at the edges of bottomlands almost as if a preordained line had been drawn. That line probably reflects moister and cooler conditions. What little charcoal is present in study pits seems to have come from elsewhere, most likely the slopes directly above the valley floors.

Wind—not old age or disease—accounts for about 80 percent of tree mortality. Pebbles flipped at random while walking along a rain forest trail would almost all hit a fallen tree trunk or a large branch: from one-tenth to one-third of the ground is covered with coarse woody debris, and in places this figure is even greater. Downed logs are so important that scientists believe their density may determine how many trees will grow in the future forest; seedling success is that strongly linked to nurse logs.

This gives particular long-term importance to studies of a stand of old-growth Douglas fir across the river from the Hoh campground. These particular trees, which are present for three miles along the lowest river terrace, indicate a past fire, wind, or flood that disrupted the previous forest; without such disturbance this species cannot become dominant here. Typically in the rain forest valleys Douglas fir grows only in relatively warm and dry sites and in the coarse soil of risers between terraces or where a raw matrix is studded with cobbles and sand. The Douglas fir across from the campground are unusually dense and therefore produce an exceptional number of windthrown logs. If nurse-log density proves to be the main determiner of tree re-

Opposite
Douglas firs are not common in the rain forest valleys, but some circumstances favor them. Such stands germinate on disturbed soil but are unable to regenerate in their own shade. They will be replaced by spruce and hemlock.

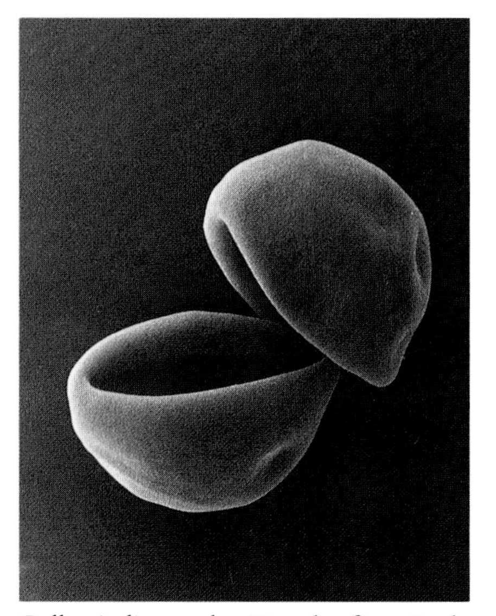

Pollen indicates that Douglas fir arrived in the Northwest about 8,000 years ago. Both it and red cedar pollen have a "deflated ball" appearance once they are shed from their cones.

generation, which many ecologists believe probable, this particular site should remain dense as the forest recycles from Douglas fir to a climax stand of spruce and hemlock. If this happens, scientists checking the trees a century or two from now will find not only a documented example of the role of disturbance but also of the role played by downed trees in determining forest density. Soil nutrients, light, and other aspects of environment probably do not govern how many trees nature grows per acre; what determines tomorrow's standing trees appears to be the quantity of today's fallen trees. Ecologists now actually speak in terms of log "recruitment" to the forest floor.

Wind is the primary recruiting agent, producing what are called blowdowns. Huge trees with stiff limbs constitute enormous surfaces for wind to push against, and the roots of even the grandest trees do not penetrate far into the earth. A fallen monarch leaves a mere dish-shaped pit only three to four feet deep and perhaps twenty feet across. No great taproot loses grip and topples the tree. On the contrary, the largest broken roots will be about as thick as a human thumb, and the largest that pull loose without breaking will be no thicker than a human thigh. Upturned, these roots form irregular vertical disks, called rootwads, which hold rocks and soil that pulled out with them. For a while ferns, herbs, and grasses continue to grow on the rootwads, though wrenched from the forest floor, and seeds arrive riding the winds and brought by birds and rodents. Such shrubs as elderberry and salmonberry, heavily browsed by elk, find their best odds for growth in the mineral soil of rootwads, out of reach. Only there are they likely to set seed and germinate seedlings.

Scattered blowdowns occur almost every year, following no particular pattern. Perhaps valley sidewalls deflect winds locally and either intensify or reduce force, perhaps not; this hypothesis has not yet been verified. On rare occasions storms that begin as tropical typhoons stray this far north, and when they contact the jet stream enormous swaths of trees may be toppled and broken. Following a storm in January 1921, for example, a forest strip thirty miles wide lay prostrate from the Columbia River mouth to Vancouver Island.

Estimates of the toll from that one storm range to eight million trees of all ages blown down and broken by the wind. Nearly two thousand blocked a single twenty-mile stretch of road northeast of Forks (Beaver to Fairholm). Old-timers and their descendants still describe the awesome crack and thud of their falling and how the ground trembled underfoot. Winds registered 132 miles per hour at the mouth of the Columbia River before their force disabled the anemometers. Forty-one years later (October 1962) wind gusting an estimated 170 miles per hour affected an even greater area and left six to twelve million downed trees. Winds of hurricane force and above have hit the coast ten times within the last two hundred years; their legacy is apparent in the age of the trees that reconquered the devastation as well as in historical records.

Paradoxically, the emerald tranquility of big-leaf maple—widely regarded as the hallmark of this forest—is also a product of sporadic

Big-leaf maples sprout in forest openings created by floods and small landslides; they need coarse soil and ample light, and so grow where tributary streams or deposits of sidewall rubble have disturbed previous vegetation.

chaos. The groves are scattered along bottomlands at the base of side-walls or where tributary streams rush to join the main rivers. Land-slides, avalanche debris, and cobbly soil washed free of fine sediments are responsible. The groves are no more than broadleaf accents in a conifer forest. They epitomize the Olympic rain forest less by being typical of its growth than by being a product of its sporadic distur-bance. For disturbance is what resets the clock here.

Human Newcomers

ARCHAEOLOGICAL EVIDENCE INDICATES THAT people have lived on the Olympic Peninsula for twelve thousand years. The finds include the remains of a mastodon butchered near Sequim at the end of the last ice age and cedar-plank houses buried by mudslide at Ozette (south of Neah Bay) close to the time that Columbus arrived in America. No one can say just when people first entered the rain forest valleys. Basalt tools dating from nine to five thousand years ago have been found at scattered sites at the base of the mountains on the north and east sides of the peninsula, and a hearth five thousand years old was discovered close to a subalpine lake in the mountains near Port Angeles. No evidence of comparable age has come from the rain for-est—which is scarcely surprising. The odds of finding early archaeo-logical sites are low at best, and the dense bottomland vegetation and rampaging rivers of the rain forest valleys further reduce chances.

Both oral traditions and early historic records indicate that Na-tive Americans had exquisitely detailed knowledge of territory and

People arrived on the Olympic Peninsula more than 10,000 years ago. This new species in the forest community had little effect on the ecosystem until after the arrival of the first European ship in the late 1770s. The Hoh Indian village (top) is situated at the river mouth, typical of settlement patterns along the entire Northwest coast. Living at river mouths gave people access to both marine and forest resources.

awareness of all its resources. Certainly by historic time Quileute, Hoh, Queets, and Quinault villagers who lived at river mouths moved up the valleys seasonally to fish and hunt elk. Elders alive today remember traveling to the camps by canoe, and they can identify traditional locations and recount which families had rights to each. Early written descriptions cover only the coastal villages, partly because European and American expeditions did not become familiar with the interior of the peninsula until the 1880s. By then epidemics had wiped out up to 90 percent of the native people, and many others, psychologically and culturally mauled, had been legislated onto reservations.

Only one party of early European explorers ventured far into the forest. These were Russian shipwreck survivors who, in 1808, spent a forlorn winter somewhere in the lower Hoh Valley and reported stumbling upon and raiding deserted camps. This handful of Europeans, accompanied by Aleut hunters and boatmen, was in the vanguard of change. Their mission had been to find a place suitable for a Russian colony, but misfortune ended that aspiration and initiated events that ultimately brought American homesteaders. Human presence, once a factor less important to the forest than elk or fungi, was then transformed into an agent of disruption as great as the ice ages themselves.

Settlers came to the rain forest to farm, beginning in the 1890s. Endless work was the only way to hold a homestead, but hopes ran high. Five years of living on the land and cultivating it brought title to 160 acres, or a few dollars per acre would buy parcels too heavily tim-

A 2,000-year-old stone carving recovered archaeologically from the coast south of Ozette (shown here about actual size).

Top and right
Gillnetting and cooking salmon at the Hoh River mouth.

The Wilhelm Moritz cabin remained where it was built in the 1890s, until it was washed away by the Hoh River in the 1970s. Settlers of Moritz's era brought in supplies and shipped out produce by river and trail. No road looped the peninsula until 1931.

bered to be worth tilling. Families newly arrived in America learned about land in the West and decided to take a chance. Workers at Puget Sound and Hood Canal sawmills heard about the potential of the west side of the peninsula and decided to try their luck after contacting specialists who would locate suitable homestead sites for a fee. All traveled by steamer to Pysht, Clallam Bay, or Neah Bay and from there made their way south, often by Indian dugout canoe. "Ten of us fit into that canoe," a Quinault Valley homesteader reminisced:

> It was about thirty-five feet long and five feet wide. The trip was two days in the rain from Neah Bay, and we had to tie up to a kelp bed and wait until the surf calmed before we could land. Then we went up the river, into the forest.

Other homesteaders approached their new homes by trail. "Mama must have known what lay ahead when she first stepped ashore," one of the four Hoh Valley homestead daughters of Dora and John Huelsdonk used to say:

> Pops opened her bridal trunk, pulled out her white ruffled petticoats, and wrapped the black iron pots and skillets with them. He said it wouldn't do for them to bang together in the paniers and scare the horses. Not with sixty miles of wilderness before reaching the ranch. . . . For sixteen years Mother never left the homestead.

Survey maps of the 1890s and 1900s show ranches dotting the valley bottoms every few miles. Some continued as ranches; others

LEFT
Minnie Nelson Peterson, daughter of Swedish immigrants who homesteaded on the Hoko River, married Oscar Peterson, the son of a Forks Prairie homesteader, and with him packed surveyors and tourists into the inner Olympics, then lived alone on a Hoh River ranch following his death in the 1960s.

TOP RIGHT
Marie Huelsdonk Lewis, daughter of John and Dora Huelsdonk who homesteaded the Hoh Valley, talks with daughter Marilyn and granddaughter Emmi.

BOTTOM RIGHT
John Fletcher, raised on a homestead near the mouth of the Hoh River, sits with his wife, Elizabeth Huelsdonk Fletcher, and neighbor, Minnie Peterson. Their deaths as octogenarians in the 1980s essentially ended the forest's pioneer era.

were abandoned. Homesteaders earned cash every way they could in order to pay taxes and buy staples like flour, coffee, and sugar. From the Hoh Valley they drove cattle and sheep to Port Townsend to sell, a week's trip. From Queets they drove to the river mouth, then south along the beach to Hoquiam. "We were too far from market for any crop except livestock to be practical," one old-timer used to sigh. "We had no proper transportation for anything else, but we could drive livestock—even turkeys."

The homesteaders who came to the rain forest had no choice but to live as a part of it. They intended to dominate the patches they claimed within it, but they were few in number and limited in technology and therefore fit into the vastness without greatly affecting the whole. Coyotes, raccoons, skunks, hawks, and owls preyed on their chickens. Elk raided gardens and trampled hayfields. ("One year the beggars waited till I'd cut the hay and then they ate it.") Men trapped animals for fur, shot cougars and bobcats for bounty, and exterminated the Olympic wolf, which they believed was a matter of self-defense and protection of livestock. Loss of wolves is perhaps the homesteaders' only major disruption of the forest's natural makeup. Large-scale intrusion into the ecosystem was not the homesteaders' role.

That changed when logging and the market hunting of elk began. These practices were motivated by viewing the forest less as a place to live than as a resource to use. Government foresters, cruising for timber in 1897, made the first detailed exploration of the rain forest val-

Forestland in the Hoh/Clearwater Block is managed by the state Department of Natural Resources. Highly productive, the land is used primarily to supply wood. Its cutting cycle is sometimes as short as forty years.

Loggers wielding hand tools began cutting Olympic Peninsula trees in the late 1800s. Most cedar logs—this one is 17 feet in diameter—were sawn into bolts, which then were either milled into shingles or hand-split into shakes.

The northern spotted owl, one of 600 species listed nationwide in 1991 as threatened or endangered, symbolizes the quandary of how simultaneously to serve economic and ecological needs. Contemporary society finds it difficult to comprehend whole ecosystems, but the welfare of seemingly minor components may significantly affect the entire web.

leys and adjacent terrain. Their report said that the cost of clearing the forest disqualified the area for agriculture because "no farmland . . . is worth such a price. . . . The most expensive item is getting rid of rotten and decaying wood, which never dries." For logging, however, the land would be ideal:

> [a railroad] with spurs up the numerous rivers to tap the timber belts would solve the basic problem of getting the timber out of the forest and into the economic mainstream. . . . Spruce stands heavy along the Hoh River and is very large. Logging will be very cheap. . . . [The Quinault River] has no value as a driving stream for logs as it has low banks and many sand bars; still, tramways can be built . . . and the timber can be handled by means of chutes and skid roads.

Despite these possibilities a rail line tying the peninsula to outside markets failed to materialize, and the forest was saved. Two decades later the Armistice ending World War I similarly saved the forest by chance, not decision. At the war's outbreak the Aviation Section of the Signal Corps suddenly demanded great quantities of Sitka spruce for airplane construction because of the wood's exceptional strength per unit of weight. Brigadier-General Brice P. Disque, in 1917 under orders to sail for France, was asked instead to head the unique Spruce Division of the U.S. Army.

Disque wrote that his mission was to increase monthly production of spruce immediately "from three million to ten million board feet." Supplying enough wood for aircraft was considered "one of the greatest unsolved problems of the war." By the summer of 1918 this need had brought ten thousand "spruce soldiers" to the Olympic Peninsula, and if the war had lasted longer the choice spruce of the rain forest valleys might well have ended up flying over Germany. Instead, continued federal designation—first as a forest reserve, later as a national park—protected the inner Olympics and a ring of surrounding lowlands. On national forest and commercial lands, extensive virgin forest lasted until the 1980s—one fleeting century, three human generations, for an ecosystem that developed five thousand years ago.

Conversely, in three generations we have gone from the pioneering economy of the late 1800s, to the beginning of spruce logging in the early 1900s, to the heavy cutting of the mid 1900s, to at least the start of a new, environmental viewpoint. Today's effort is to protect much of the remaining natural forest for the sake of diversity, and to apply lessons learned there to enhance and sustain production on lands managed for timber production. Respect for the ancient working of the ecosystem is beginning to be understood scientifically.

Today and Tomorrow

ECONOMY AND ECOLOGY SHARE the same Greek root, *oikos*, the word for "home." This fundamental truth formerly was a matter of daily awareness: nature's rhythms and resources governed human livelihood until a century or two ago. Increased technology brought an ability to manipulate and redistribute the environment's riches and, with it, an assumption that we live apart from

The forest has evolved over millennia, changing at its own pace in response to environment and genetic codes. Human decisions now divert established patterns and quicken the pace of change. Outcomes are not yet known.

nature. People began to view forests not as a natural community but as a commodity. Our species has always used and altered its environment, but the ways of doing so previously were simple, human population was low, and the attitude toward nature was respectful. Of the three, attitude then made the least difference to the forest, but it is now the most important. What we care about, we care for.

E. O. Wilson, Harvard ecologist, points out that a single handful of soil and organic litter scooped from a forest floor holds "more order and richness of structure, and particularity of history, than the entire surface of all the other planets combined." The web of life within the Northwest's forests exemplifies the point. With the extent of its complexity now beginning to be understood, it is no wonder the need to continue meeting economic demand sends corporations and agencies to the studies of ecologists. To have trees they also must manage unseen realities from mycorrhizal fungi to canopy scuzz. No wonder, too, that ecologists are humble in their knowledge. Nature has been honing its lesson plans a long time, and we have only begun to study.

What do we find? Endless surprises. Even viewed as treasure for our benefit, genetic diversity per se has incalculable value. Instead of limiting food crops to a few basics worldwide—wheat, rice, corn, peas, beans—we could draw more fully from the estimated 7,000 species of plants people are known to have eaten through the years. For oil, we could cultivate more Babassu palms, an Amazonian species, capable of producing an impressive fifty barrels of vegetable oil per acre from the coconutlike fruits of just two hundred trees. For preventing brain damage from stroke, a drug found in spider venom may prove useful. In the late 1980s, a compound called taxol, which is present in yew trees, led to the development of a new treatment for cancer. Previously forest managers had regarded yews as weeds. They grow scattered and take a century to reach a foot or so in diameter, and rarely stand more than about 50 feet tall. Even so, Native Americans held them in high regard. They knew the wood's unique bending strength, hardness, and resistance to shock, and they used it for bows, digging sticks, whaling harpoon shafts, and clubs for striking and killing wounded sea mammals and fish. They also made medicines (including a contraceptive) from yew bark, berries, and needles. But for modern society, the trees seemed a nuisance until discovery of the existence—and value—of this one chemical, which halts malignant growth by encasing cells with a fibrous substance.

At first, a single pound of taxol required the bark of 2,000 yew trees and thereby created a quandary. Should all possible yews, in whatever forest, be stripped for the treatment of current cancer patients? Or should some be reserved for future patients, or for whatever other useful substances the species may also have? Is it wise to eradicate genes or should some be safeguarded for their own sake?

Today, the question with regard to taxol is mute. Wild yews no longer are needed for the drug because they have been replaced by trees specifically cultivated for the purpose and also by taxol-like chemicals produced in laboratories. Even so, the example of taxol points to the

The bald eagle is part of the diverse ecosystem of the Bogachiel, Hoh, Queets, and Quinault valleys, which rank among the magnificent wild communities of Earth. A 1991 Forest Service research report concludes that such ecosystems are not only more complex than we think, they are more complex than we *can* think. Each addition to the existing knowledge of them is a step toward true understanding of the whole.

role of diversity per se as a repository and library of still unrecognized—or forgotten—benefit to humans. With computers now common, we know the principles of thinking in terms of "bits," an ordering of information into pairs of equally likely alternatives. On this basis Wilson suggests that a single bacterium possess ten million bits of genetic information; a fungus, one billion bits; and an insect, as many as ten billion bits, depending on species. This is what conservationist Aldo Leopold had in mind when he pointed out that the first rule of successful tinkering is not to throw away the parts—not *any* of the parts. Should we think short-range or long-range?

The problem with extinction is that it eliminates parts, often before we even know what they are. The current rate of loss differs radically from what it was only four centuries ago. It is estimated that one extinction occurred per thousand years before 1600; a thousand extinctions occur each year now, mostly in the tropics. Humans are the chief cause, the first time ever that one species has so affected the community.

We can plant trees but not forests. They make themselves. Their complexities are beyond our awareness and knowledge; we have only begun to hear their elemental whisper.

Glossary

Acre. 4,840 square yards, originally the amount of land an ox could plow in a day.

Alluvium. Sediments, from silt to rocks, carried and deposited by running water.

Anadromous. Refers to the life cycle of certain fish such as salmon and steelhead, which mature in the ocean but spawn in fresh water.

Bark. All tissues outside the cambium.

Biodiversity. The variety of species living within an area.

Biomass. The total quantity of organisms living at a given time within a unit of space.

Biome. The type of community produced by organisms living together.

Blowdown. An expanse—sometimes several acres—of trees toppled and broken by wind.

Broadleaf trees. Those with leaves as opposed to needles (which technically are one type of leaf).

Cambium. Layer of cells immediately under the bark, where the annual growth of wood and bark takes place.

Canopy. The combined crowns of trees and the resulting structure of their branches, twigs, and leaves.

Cellulose. The chief cell-wall component of most plants, which accounts for their solid framework.

Clear-cut. A single cutting of all trees in an area, which will be replaced by a new, even-aged stand, often of a single species.

Climax. Culminating stage of plant succession, which will essentially perpetuate itself unless the environment changes.

Colonize. The process by which species move into a new habitat and establish themselves.

Commodity. Any resource product that is an article of commerce.

Competition. The effect of multiple organisms depending on a limited amount of food or space.

Conifer. Trees that bear cones; most are evergreen; all belong to *Gymnospermae*, an earlier and more primitive order than *Angiospermae*, the flowering plants.

Deciduous. Trees and shrubs that shed and regrow leaves every year, as opposed to evergreens, which retain foliage year-round.

Dominant. The most abundant or influential species within a community.

Ecology. The science of relationships between organisms and their environment.

Ecosystem. A community of organisms together with their environment and the interactions between these components.

Edge. The contact point between two or more distinctive ecosystems.

Endangered species. Any species in danger of extinction within a significant portion of its range or its entire range.

Epiphyte. A plant growing non-parasitically on another plant.

Evolution. The process by which successive generations of species and groups of organisms develop modifications.

Extinction. The dying-out of a species, sometimes referring to the total existence of the species, sometimes to its loss within a limited area (also called extirpation).

Flagella (sing., flagellum). Whiplike appendages that make certain cells and organisms motile.

Fruiting body. The seed- or spore-bearing organ of a plant.

Genus. A group of species so closely related as to be genetically distinct from other groups.

Habitat. The natural or normal place where an organism lives.

Hardpan. An impermeable layer within the ground caused by iron cementing soil particles together.

Herb. A seed-producing, non-woody plant.

Hyphae (sing., hypha). Threadlike structures that constitute the body of a fungus.

Hyporheic zone. A gravel zone contiguous with a river channel and hydraulically interactive with it.

Indicator species. A plant or animal species so closely linked to a particular type or stage of community development that its presence suggests the nature of the rest of the community.

Invertebrate. A term referring to animals without backbones, e.g., insects, spiders, slugs, snails, worms, etc.

Lichen. The merging of algae and fungi, and sometimes cyanobacteria, into a distinct organism with one of three possible forms: crustose (a thin encrustation on rock or wood); foliose (leaflike); fruticose (with upright stalks, usually branched).

Lignin. The main non-cellulose constituent of wood.

Litterfall. The rain of organic debris and accumulated minerals through the crown of a tree or shrub.

Logjam. Woody debris, especially logs and large branches, caught on the gravel bar or bank of a river.

Mature. Refers to the stage at which the mean annual production of the community has culminated.

Metamorphosis. Change in form from one life stage to the next, as in insects.

Mushroom. The aboveground, spore-bearing portion of some fleshy fungi.

Mycelia (sing., mycelium). Collective term for fungal hyphae.

Mycorrhizae. Associations of fungal cells with the root systems of vascular plants, which permit an exchange of nutrients, minerals, and water.

Nitrogen fixation. The conversion of gaseous nitrogen into a form that is useful to plants and animals.

Nurse log. A fallen tree that in the process of decay becomes a reservoir of nutrients and moisture and a habitat for mosses and seedlings.

Old growth. Forests that have passed maturity and typically have a diverse composition and structure.

Phloem. The softer part of woody tissue. (*See* xylem.)

Photosynthesis. The process by which plants use light to convert carbon dioxide and water into carbohydrates.

Plantation forest. Trees planted and managed for commercial production.

Regeneration. Successful establishment of the seedlings and saplings of a particular plant species.

Rhizome. The underground stem of a vascular plant.

Rootwad. The intact root mass of a fallen tree together with the soil and rock clinging to it.

Sapling. Young trees beyond the seedling stage, generally a few feet high and an inch or so in diameter at mid-height.

Second growth. Natural regrowth following drastic disturbance, whether natural or the result of logging.

Seedling. Newly germinated, or young, vascular plants.

Snag. A standing dead tree.

Species. Closely related individuals having certain permanent and characteristic traits that distinguish them from all other groups.

Spore. The usually unicellular reproductive body of certain organisms such as fungi.

Stand. An aggregation of trees within a particular area, sufficiently uniform in age and species to be distinguishable from adjoining forest.

Succession. The stages within a plant community from bare ground to climax.

Threatened species. Plant or animal species likely to become endangered within the foreseeable future.

Truffles. Fungi living entirely underground.

Ungulate. Taxonomic order of mammals with legs ending in hooves.

Windthrow. A tree toppled by wind.

Xylem. Collective name for cells, vessels, and fibers of woody tissue.

Reading List

Alt, D.D., and D.W. Hyndman. *Roadside Geology of Washington.* Mountain Press Publishing Co., Missoula, MT, 1984. A road guide for the entire state including a chapter on the Olympic Peninsula.

Arno, Steven F., and Ramona P. Hammerly. *Northwest Trees.* The Mountaineers, Seattle, 1977. Superbly illustrated guide to identification and basic characteristics of tree species.

Brown, E. Reade, tech. ed. *Management of Wildlife and Fish Habitats in Forests of Western Oregon and Washington,* Part 1: Chapter Narratives. U.S. Dept. of Agriculture, Forest Service, Portland, OR, 1985. Chapters deal with subjects such as plant communities, riparian ecosystems, salmonids, deer and elk, silviculture options, snags, and dead and downed woody material.

Cahalane, V.H. *Mammals of North America.* Macmillan Co., New York and London, 1961. Informative and enjoyable introduction to mammals, including species native to the Olympic Peninsula.

Dodwell, Arthur, and Theodore F. Rixon. *Forest Conditions in the Olympic Forest Reserve, Washington*; U.S. Geological Survey, Professional Paper No. 7, 1902. Not intended for general readers, but interesting as the earliest detailed description of the essentially primeval forest.

Ervin, Keith. *Fragile Majesty: The Battle for North America's Last Great Forest.* The Mountaineers, Seattle, 1989. Summation of natural history and politics of old-growth forests in the Northwest.

Fellows, Larry A., proj. leader. *Spotted Owl Guidelines,* vol. 1. U.S. Dept. of Agriculture, Forest Service, Portland, OR, 1988. Summarizes research and management alternatives.

Fletcher, Elizabeth Huelsdonk. *The Iron Man of the Hoh.* Creative Communications, Port Angeles, WA, 1979. Fascinating family account of the first pioneers to homestead in the Hoh Valley.

Franklin, Jerry F., Kermit Cromack, Jr., William Denison, Arthur McKee, Chris Maser, James Sedell, Fred Swanson, and Glen Juday. *Ecological Characteristics of Old-Growth Douglas-fir Forests.* Gen. Tech. Report PNW-118, U.S. Dept of Agriculture, Forest Service, Portland, OR, 1981. Readable and informative summation of Northwest old-growth ecology.

Franklin, Jerry F., and C.T. Dyrness. *Natural Vegetation of Oregon and Washington.* Oregon State University Press, Corvallis, 1988 (originally published in 1972 by the U.S. Forest Service). Standard reference summarizing Northwest vegetation by community type and overall ecological principles; extensive bibliography.

Harris, Larry D. *The Fragmented Forest; Island Biogeographic Theory and the Preservation of Biotic Diversity.* Univ. of Chicago Press, 1984. Excellent summation of the value of diversity and current threats to it.

Henderson, Jan A., David H. Peter, Robin D. Lesher, and David C. Shaw. *Forested Plant Associations of the Olympic National Forest.* R6 Tech. Paper 001-88. U.S. Dept. of Agriculture, Forest Service, Portland, OR, 1989. Extensive survey of Olympic Peninsula ecosystems and research including sections on geology, climate, soils, fauna, and human history as well as vegetation.

Hitchcock, C. Leo, and Arthur Cronquist. *Flora of the Pacific Northwest.* Univ. of Washington Press, Seattle, 1973. Guide to plant identification, well illustrated with line drawings.

Kirk, Ruth, with Richard D. Daugherty. *Exploring Washington Archaeology.* Univ. of Washington Press, Seattle, 1978. Overview of prehistory, including description of western Olympic Peninsula archaeology.

Kirk, Ruth. *Tradition and Change on the Northwest Coast.* Univ. of Washington Press, Seattle; Douglas & McIntyre, Vancouver and Toronto, 1986. Summation of ethnology and history of three Native American groups on the southern Northwest Coast.

Kirk, Ruth, and Carmela Alexander. *Exploring Washington's Past: A Road Guide to History.* Univ. of Washington Press, Seattle, 1991. Human history of the state with the Olympic Peninsula presented as one of eight travel regions.

Kruckeberg, Arthur R. *The Natural History of Puget Sound Country.* Univ. of Washington Press, Seattle, 1991. Beautifully produced introduction to regional natural history with application to the rain forest although dealing specifically with the Puget Sound area; abundantly illustrated.

Kozloff, Eugene N. *Plants and Animals of the Pacific Northwest: An Illustrated Guide to the Natural History of Western Oregon, Washington, and British Columbia.* Univ. of Washington Press, Seattle, 1976. Identification and description of common Northwest species.

Lien, Carsten. *Olympic Battleground.* Sierra Club, San Francisco, 1991. Convincingly written, discouraging account of the politics of preservation in Olympic National Park.

Maser, Chris, and James M. Trappe, eds. *The Seen and Unseen World of the Fallen Tree.* Gen. Tech. Report PNW-164, U.S. Dept. of Agriculture, Forest Service, and U.S. Dept. of the Interior, Bureau of Land Management, Portland, OR, 1984. Detailed, readable description of the ongoing role of trees following death, including nutrient cycling and biotic succession.

Maser, Chris, Robert F. Tarrant, James M. Trappe, and Jerry F. Franklin, tech. eds. *From the Forest to the Sea: A Story of Fallen Trees.* Gen. Tech. Report PNW-GTR-229, U.S. Dept. of Agriculture, Forest Service, Portland, OR, 1988. Summarizes research concerning large trees on the forest floor, in rivers and streams, in estuaries, at the beach, and in the ocean, together with forestry options for public lands.

McKee, Bates. *Cascadia: The Geologic Evolution of the Pacific Northwest.* McGraw-Hill Book Co., New York, 1972. Outstanding summary providing overall geologic context of the entire state together with specific description of the Olympic Peninsula.

McKenny, M., and D.E. Stuntz. *The Savory Wild Mushroom.* Univ. of Washington Press, Seattle, 1971. (Revised and enlarged by Joseph Ammirati, 1987.) Field identification guide to finding and eating mushrooms.

Moir, William H. *Forests of Mount Rainier National Park: a Natural History.* Pacific Northwest Interpretive Association, Seattle, 1989. Enjoyable account of field observations and research; generally applicable to Olympic Peninsula forests, although it describes those at Mount Rainier.

Nadkarni, Nalini. "Roots that Go Out on a Limb," *Natural History,* vol. 94, no. 2 (Feb. 1985). Illustrated article on Hoh Valley epiphyte research.

Norse, Elliott A. *Ancient Forests of the Pacific Northwest.* Island Press, Washington, D.C., 1990. Excellent non-technical, but informative and detailed, presentation of Northwest forest ecology.

Northwest Environmental Journal, vol. 6, no. 2 (Fall/Winter). Institute for Environmental Studies, Univ. of Washington, Seattle, 1990. Compilation of articles by Northwest old-growth researchers; somewhat technical.

Post, Austin, and Edward LaChappelle. *Glacier Ice.* The Mountaineers/Univ. of Washington Press, Seattle, 1971. Spectacularly illustrated and well-written general introduction to glaciation in the Northwest.

Ruggiero, Leonard F., Keith B. Aubry, Andrew B. Carey, and Mark H. Huff. *Wildlife and Vegetation of Unmanaged Douglas-fir Forests.* U.S. Dept. of Agriculture, Forest Service, Portland, OR, 1991. Professional report summarizing current research on the Northwest environmental setting, flora, and fauna.

Tabor, Roland W. *Guide to the Geology of Olympic National Park.* Univ. of Washington Press, Seattle, 1975. Popular guide to peninsula geology.

VanPelt, Robert. *Washington Big Tree Program.* College of Forest Resources, Univ. of Washington, Seattle, 1991. Official listing of record-size trees for the entire state.

Van Syckle, Edwin. *They Tried to Cut It All: Grays Harbor—Turbulent Years of Greed and Greatness.* Pacific Search Press, Seattle, 1981. Lively logging and milling history of the southern base of the Olympic Peninsula.

Whitney, Stephen. *Western Forests.* The Audubon Society Nature Guide Series, New York, 1988. Illustrated guide to forest natural history.

Index

Bold numbers refer to photographs or illustrations; *italic* numbers refer to captions.

All photographs by Ruth and Louis Kirk except as follows:

Kim Ziegler, cover; Ross Hamilton, 1; U.S. Army Corps of Engineers, 20–21; Janis Burger, 36 bottom, 46, 48 bottom right, 51, 58, 72 center right, 81, 95, 119; Larry Jones, U.S. Department of Agriculture, Forestry Sciences Laboratory, Olympia, WA, 50, 53, 77 bottom; Fred Rhoades, 56, 57 top left and right; George Carroll, 57 bottom left, 59 bottom; Tom and Pat Leeson, 66, 77 top; Ed Schreiner, National Park Service, Olympic National Park, 67 bottom left; Olympic National Park, 78 left bottom; Tom Odell, U.S. Department of Agriculture, Forestry Sciences Laboratory, Corvallis, OR, 86; R. A. Wood, 89; Jim Grace, 90 top; B. Zak, 90 bottom left; Jim Trappe, 90 bottom right, 91 bottom; Gary A. Laursen, 91 top; Douglas Houston, National Park Service, Olympic National Park, 101 right; John Owens, Department of Biology, University of Victoria, British Columbia, 108, 110 bottom; Jerry Franklin, 118 left; Washington State Historical Society, 118 right.

All drawings are by Gretchen Bracher and include the following adaptations:

p. 49 is adapted from Franklin, J. F., T. Spies, D. A. Perry, M. Harmon, and A. McKee, "Modifying Douglas-fir Management Regimes for Non-timber Objectives," *Douglas-fir: Stand Management for the Future*, edited by C. D. Oliver. College of Forest Resources, Univ. of Washington, Seattle, 1986 (p. 375).

p. 56 is adapted from Franklin, J. F., Kermit Cromack, et al. *Ecological Characteristics of Old-Growth Douglas-fir Forests* (p. 21). General Technical Report PNW-118, U.S. Dept. of Agriculture, Forest Service, 1981.

p. 75, top, is adapted from Maser, Chris, and James M. Trappe, eds. *The Seen and Unseen World of the Fallen Tree* (p. 3). General Technical Report PNW-164, U.S. Dept. of Agriculture, Forest Service, and U.S. Dept. of the Interior, Bur. of Land Management, 1984.

p. 75, bottom, is adapted from Franklin, J. F., and Richard H. Waring, "Distinctive Features of the Northwestern Coniferous Forest: Development, Structure, and Function," in Richard H. Waring, ed., *Forests: Fresh Perspectives for Ecosystem Analysis*, Proceedings of the 40th Annual Biology Colloquium (p. 63), Oregon State University Press, Corvallis, 1980.

p. 97, Ward, J. V., and J. A. Stanford, "Groundwater Animals of Alluvial River Systems: A Potential Management Tool," in N. S. Greig, ed., *Proceedings of the Colorado Water Engineering and Management Conference* (p. 394), Colorado Water Resources Institute, Fort Collins, 1989.

p. 99, Triska, Frank J., and Kermit Cromack, Jr. "The Role of Wood Debris in Forests and Streams," in Richard H. Waring, ed., *Forests: Fresh Perspectives for Ecosystem Analysis*, Proceedings of the 40th Annual Biology Colloquium (p. 180), Oregon State University Press, Corvallis, 1980.

Map pp. 20–21 uses a U.S. Army Corps of Engineers radar photograph as base.